12/6/13

Dann,
may our passion
and respect for
great design and
heritage continue
to burn bright.
Cameron

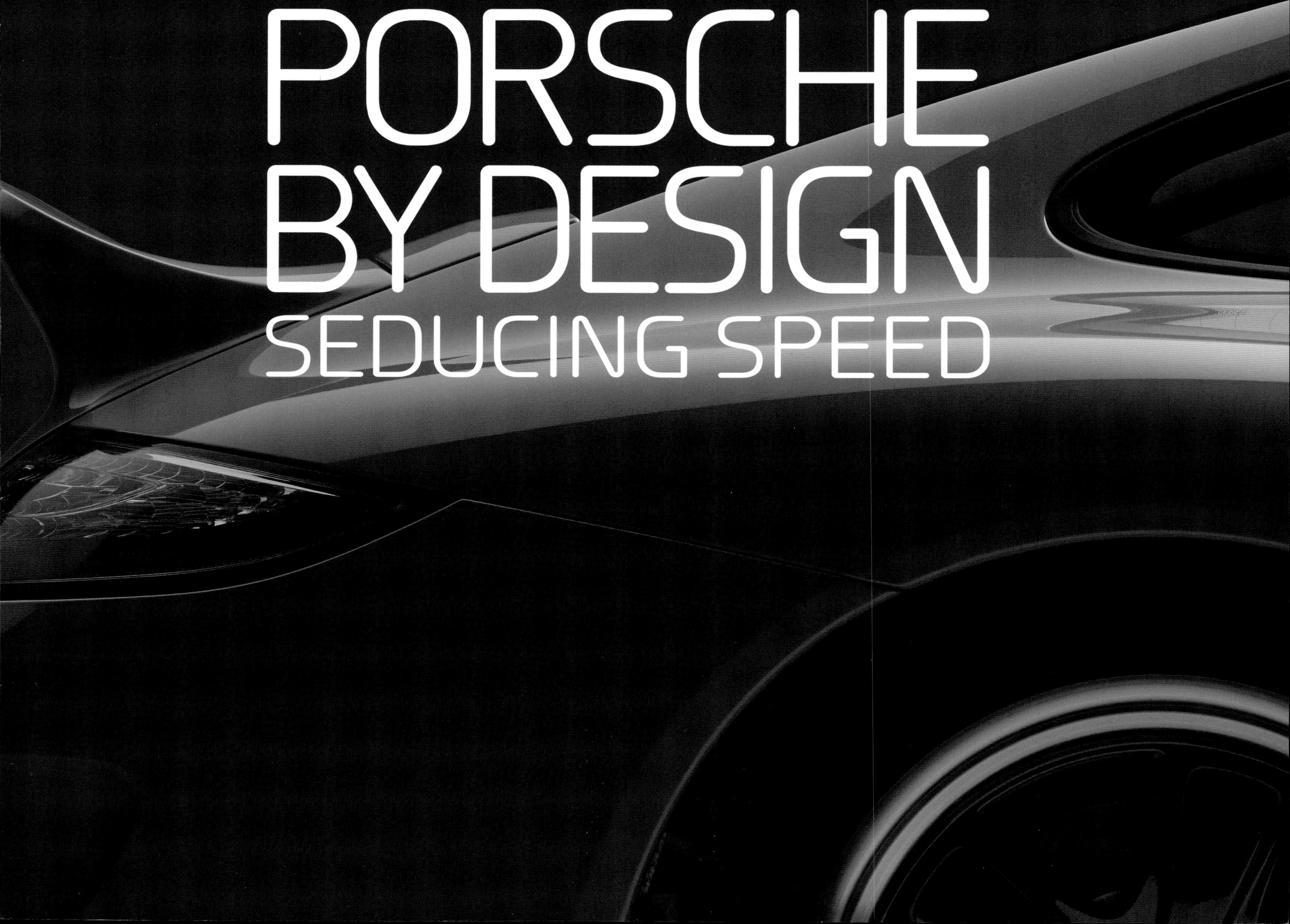

PORSCHE BY DESIGN

SEDUCING SPEED

EDITED BY KEN GROSS

WITH BARBARA WIEDEMANN

PORSCHE BY DESIGN
SEDUCING SPEED

north carolina museum of art

Edited by Ken Gross, Barbara Wiedemann, and Laurie Evans
Indexed by David Luljak
Designed and typeset by Barbara Wiedemann
Photo editor Barbara Wiedemann

Cover: **Porsche Type 64 Berlin-Rom Racer**, 1938,
Collection of the Automuseum Prototyp, Hamburg,
Germany, Photograph © 2013 Photographers Hamburg

Frontispieces: **Porsche Type 911 Sport Classic Carrera** (detail), 2010,
Courtesy of the Ingram Collection, Photograph © 2013 Michael Furman;
Porsche Type 804 Formula One, 1962, The Porsche Collection of
Ranson W. Webster, Photograph © 2013 Peter Harholdt

Library of Congress Control Number: 2013944444

Printed in Korea, through Four Colour Print Group, Louisville, Kentucky

North Carolina Museum of Art
2110 Blue Ridge Road
Raleigh, North Carolina
www.ncartmuseum.org

The North Carolina Museum of Art, Lawrence J. Wheeler, director, is an agency of the North Carolina Department of Cultural Resources, Susan W. Kluttz, secretary. Operating support is provided through state appropriations and generous contributions from individuals, foundations, and businesses.

ISBN 978-0-88259-901-4

Special thanks to:

PORSCHE

CONTENTS

DIRECTOR'S FOREWORD

Lawrence J. Wheeler
Director, North Carolina Museum of Art

I am delighted to introduce this book, a collection of personal perspectives on Porsche, to accompany *Porsche by Design: Seducing Speed*, the first automotive design exhibition presented by the North Carolina Museum of Art. We could not have assembled such a noteworthy exhibition without the expertise and professionalism of guest curator Ken Gross, a renowned automotive journalist, museum consultant, and former director of the Petersen Automotive Museum in Los Angeles. From the beginning the depth and breadth of Ken's automotive knowledge have helped to shape this collection of Porsche automobiles. Ken helped us bring together superlative examples of the signature Porsche design principles, illustrating the entire history of this unique manufacturer with breathtaking examples of minimalism, aerodynamic fluidity, and technical innovation.

I am pleased that managing curator Barbara Wiedemann, who also serves as the Museum's director of publications, could support Ken's efforts with such creativity and dedication, and I am charmed by the knowledgeable voices that Ken and Barbara have brought together in this book to tell the Porsche story. The distinctive viewpoints of the thirteen authors—from the driver's seat of a 917 at Le Mans to the helm of the design offices in Zuffenhausen—bring the story of Porsche design alive.

As we at the Museum travel down this new road, I've found myself intrigued by the relentless perfectionism inherent in every variation of the now fifty-year-old Porsche 911. I've noted the quiet aerodynamic minimalism of the 1938-era Type 64, a precursor to all of the other automobiles in this exhibition. And I have been captivated by the elegant details of each of these twenty-two artfully restrained Porsche cars. Are cars art? I leave that to the visitor to decide. As curator Arthur Drexler found in 1951 at the Museum of Modern Art, automobiles can be "hollow, rolling sculpture." I have no doubt that these particular finely crafted machines deserve our attention as exemplars of the highest form of design achievement.

Special thanks go to a generous group of lenders from around the country and, indeed, the world for sharing their passion for Porsche in such a generous way. We thank Porsche Cars North America for its support of the exhibition by underwriting our *Porsche by Design* short-form video series and providing access to the historic archives at the Porsche Museum in Zuffenhausen, Germany. Even more important, PCNA facilitated the loan of four significant automobiles from the Porsche Museum to our exhibition in Raleigh. Generous financial support from Jiffy Lube and Performance Porsche helped us to bring this evolution of the Porsche design aesthetic to North Carolina. We are thankful for the additional funding provided by steadfast Museum supporters Quintiles and for the generosity of Peter Millar and Wells Fargo, The Private Bank. Capitol Broadcasting Company continues to play a significant role in bringing the arts to North Carolina as our in-kind media partner. And I am indebted to the North Carolina Museum of Art staff for taking on a new challenge with a rigorous professionalism that sets a new bar for special exhibitions at the Museum.■

ACKNOWLEDGMENTS

Ken Gross
Guest Curator

The creation of *Porsche by Design: Seducing Speed*, like all notable exhibitions and catalogues, involved inspired effort from a talented, enthusiastic, and diverse collection of people.

First and foremost, we must acknowledge the accomplishments of the remarkable Porsche family, from Dr. Ferdinand Porsche through his son Ferdinand ("Ferry") Porsche, to Ferry's son Ferdinand Alexander (F. A. "Butzi") Porsche, and to the Piëchs, Louise and Ferdinand. They clearly established a remarkable automotive tradition and have left an undying legacy. Porsche cars enthrall and reward all who are fortunate enough to drive them. Porsches continually compete against the world's best—and win. The continuity of the family's involvement; the unbroken visual lineage of the cars themselves; the design and engineering innovation and mechanical excellence; and, of course, the people who created the cars and raced them inspired this exhibition.

Barbara Wiedemann of the North Carolina Museum of Art has been the catalyst, driving force, organizer, visionary, occasional nag, in the best of ways, talented scribe, and principal book designer. Barbara has lived this project from its inception. Her dedication, energy, and creativity contributed immensely to its success. I know she would like to recognize Raleigh's own Bob Irwin and Bill Hopkins, friends and early visionaries for this Museum's exhibition.

An exhibition is only as good as its lenders and their loans. And we have had most generous contributors. Attempting to tell more than eight decades of a company's history with just twenty-two cars would have been an impossible task were it not for them.

It begins with Porsche AG in Stuttgart. We'd like to thank Rolf Sprenger for being an early supporter and ambassador to the Porsche Museum for us, and for sharing his insights on camera as well. Achim Stejskal, the director of the Porsche Museum, hosted us at the museum's elegant Christophorus Restaurant and noted, "You only want our best cars," then graciously responded to our requests. Klaus Bischoff's behind-the-scenes encouragement and influence were priceless. Manager of Historic Archives Dieter Landenberger, the Porsche Museum's archivist extraordinaire and a star of our video series, and his colleagues Jens Torner, Dieter Gross, Tobias Mauler, and Yvonne Knotek tirelessly helped us find just the right photos and film clips for the video series, the exhibition labels, and the book. Astrid Böttinger and Katja Leinweber, also from the Porsche Museum, unflaggingly supported our projects. Alexander Klein coordinated four very special loans from the Porsche Museum for us.

We want to sincerely thank and acknowledge these additional contributing lenders and their associates: Thomas König and his colleague Oliver Schmidt from the Automuseum Prototyp in Hamburg, Germany; Diane and Don Meluzio; Miles C. Collier and Scott George from the Collier Collection and The Revs Institute, who helped out on-camera as well; Chad McQueen; William E. "Chip" Connor II and his associate Susan Skidmore; Ralph Lauren and his associates Mark Reinwald and Rebecca Evans; lender Larry Bowman along with Jamie Karrick of Bowman Motors; Ranson W. Webster and his collection manager, Bart Galloway; the Joplin family and Jeff Jampol, along with Greg Harris, from the Rock and Roll Hall of Fame and Museum; and Cameron Healy, who with his wife, Susan Snow, lent two cars; he also wrote an essay about racing them.

The Ingram family—Robert A. Ingram, his effervescent wife, Jeanie, and their knowledgeable sons Cameron and Rory—assisted with every aspect of this exhibition as enthusiastic cheerleaders, generous lenders, genial hosts for special events, and, in Cam's case, as essayist and on-camera talent. They exemplify the adage that great collectors share their collections. We can't thank them enough.

This exhibition focuses on Porsche engineering and design, past and present. We thank Porsche designers Grant Larson and Anthony "Tony" Hatter for their on-camera insights; also former Porsche design chief Harm Lagaaij, for his contributions to the video; and Michael Mauer, current head of Style Porsche, for his insightful essay. Hans Mezger, brilliant Porsche engineer, also provided valuable insights both on camera and off.

We simply could not have accomplished our task without Bernd Harling from Porsche Cars North America. Bernd's intimate knowledge of the marque and its people and his fine reputation with the Porsche AG staff, along with his keen wit, ensured we had the company's support to accomplish our task. We'd also like to thank PCNA's Detlev von Platen, Michael Bartsch (recently named CEO of Porsche Cars Australia), Thierry Kartochian, Andre Oosthuizen, Steve Janisse, and Nick Twork.

Kevin Heimbaugh, Ralph Moore, and Tom Roos, enthusiastic members of the Porsche Club of America, supported and encouraged this project from the outset, as did Vu Nguyen, PCA executive director, and Pete Stout, editor in chief of Porsche *Panorama* and director of integrated communications at PCA, who have helped ensure the exhibition was well publicized. The exhibition materials, including this book, visually sparkle because of the work of photographers Michael Furman, Peter Harholdt, and videographer Art Howard. Connie Nyholm and Kerrigan Smith provided track time at the lovely Virginia International Raceway (VIR) for special footage of some historic Porsche cars in motion.

Porsche authority Karl Ludvigsen wrote an elucidating essay and shared his wisdom via our audio tour. Robert Cumberford, a world-renowned design analyst, generously contributed his expertise to the video and audio tour and wrote a fine essay for this book to boot. Five-time Le Mans and three-time World Sportscar Champion Derek Bell was twice gracious in that he shared his insights on camera and in writing; as were the *Wall Street Journal*'s irrepressible Dan Neil and champion race driver/filmmaker Jeff Zwart. We were fortunate to have experts Randy Leffingwell, Pete Lyons, and Denise McCluggage write so poetically for this book. My friend Jay Leno helped us kick off the video series from his "garage" on the West Coast. And Webb Farrer helped us ensure that the installation went smoothly.

Thanks go to the North Carolina Museum of Art's insightful Director Lawrence J. Wheeler and to Deputy Director and CFO Caterri Woodrum for their support. Thanks also to Exhibitions Manager Tiara Paris; Chief Registrar Maggie Gregory and her colleagues; and Curator of Design Eric Gaard and his exhibition design team, who provided the framework for this exhibition under the guidance of Dan Gottlieb, director of planning and design. Jennifer Bahus and her Marketing and Communications group enthusiastically supported this exhibition; as did Director of Development Ellen Stone and her staff. My thanks go to Director of Performing Arts and Film George Holt and his crew for their efforts to bring our theme to life via the Museum's film series leading up to and during the exhibition; Sandy Rusak's direction regarding educational programming endeavors has been notable as well. And my special thanks go to editor Laurie Evans from the Publications team, who polished our words and supported our research for this fine book and the exhibition materials.

I have been delighted to serve as guest curator and a contributor to this book and the videos. As a former Porsche owner who still dreams about his '61 356B S90 coupe, I have always admired the marque. Helping to create this exhibition has been a wonderful opportunity to become immersed in the history of Porsche—the cars, the family, and a remarkable company. ■

INTRODUCTION

Ken Gross
Guest Curator
Hamilton, Virginia
June 2013

It was a dream assignment to help curate *Porsche by Design: Seducing Speed*. I'll admit it. I love Porsche cars. Back in college my friend Jeff Hemmings let me drive his 356A on twisty back roads down to Watkins Glen; I was totally hooked. Afterward I yearned for a Porsche, and as soon as I finished graduate school, I bought a preowned '61 Super 90 coupe. With its raucous Bursch exhaust, it was loud, quick, everything I thought it would be, and much more. I've coveted Porsches ever since, and although I don't own one at the moment, I console myself with an extensive Porsche library and frequent perusals of the classifieds.

I'm aware that a great deal has been written about Porsche over the years, so when managing curator—and this book's meticulous designer—Barbara Wiedemann and I began planning the catalogue, we wanted something a bit different. In addition to presenting and illustrating each of the twenty-two cars on display in the exhibition, we elected to tell key elements of the Porsche story, expressed by talented men and women who had designed, engineered, and raced these exceptional cars. We also gathered recollections from Porsche historians, restorers, and collectors, along with tributes from noted writers.

To accomplish this we assembled an all-star cast. Karl Ludvigsen, the premier Porsche historian and the author of the three-volume masterpiece *Excellence Was Expected*, was surely a "must-have" contributor, as was the incomparable Robert Cumberford, whose insightful design analysis helps us to appreciate the timeless shape that Ferdinand Porsche and Erwin Komenda established from the beginning. Miles Collier, articulate and keenly observant, a top-shelf Porsche collector/racer and a lender to this exhibition, makes a cogent case for preservation. Cameron Healy enthusiastically takes us onto the track in his historic racing 904/6. Derek Bell, courageous and modest, makes piloting the wickedly fast 917K sound easy.

Denise McCluggage charms us with wonderful reminiscences of a simpler era, driving her very competitive Porsche Spyder, and Dan Neil's insights into Porsche's beginnings, coupled with his exciting contemporary GT3 R ride, link the yin and yang of Porschedom. Pete Lyons's vivid Can-Am memories recall those Mark Donohue days of blindingly quick 917/10s and 917/30s. Pete was there, and no one tells it better. Cam Ingram's award-winning restoration shop was chosen by Hans-Peter Porsche to bring back a Pebble Beach–winning Gmünd coupe. Nuff said. Michael Mauer explains the delicate balance of designing new Porsche cars while remembering old ones, and then moving bravely forward. Jeff Zwart and Randy Leffingwell are articulate photographers and vintage racers. We were fortunate to have everyone's help.

Photographic talent abounds in this catalogue, with the exquisite work of Michael Furman and Peter Harholdt and that of our well-traveled photographer/videographer Art Howard. Each takes a different and expressive approach to his craft. The remarkable images and posters from the comprehensive Porsche Museum Archives in Stuttgart were provided by Jens Torner and Dieter Landenberger. The beauty of this presentation is how well it all works together. Barbara Wiedemann's artistic touch in the fine design of this book, along with her studied involvement in so many elements of this exhibition, makes the result very special.

Genug! It's time to enjoy the words and images.

But before you do, one last thought. Like no other car company, like no other automotive dynasty, Porsche has a history of perfect continuity of familial talent—you could say genius—working closely with dedicated engineers, designers, technicians, and racers, all understanding that there is nothing like a Porsche, now and forever.■

AUTO UNION

KARL LUDVIGSEN

PORSCHE ENGINEERING: THE BEGINNING

From his twenty-fifth year, Ferdinand Porsche enjoyed a brilliant career in the motor industry. At that age in 1900 he joined Vienna's Lohner as chief engineer to produce electric-drive vehicles of rare virtuosity. In 1907 entrepreneur Emil Jellinek tapped Porsche as technical director of automaker Austro-Daimler. As managing director from 1913, Porsche led Austro-Daimler through the war years. Named director of engineering in 1923 at Stuttgart's Daimler, fabled Mercedes maker, Porsche oversaw its merger with Benz before leaving in 1928 after five fruitful years. He designed, among other cars, the vaunted Mercedes Model S, SSK, and SSKL.

LEFT: 1937, Ferdinand Porsche in discussion with driver Bernd Rosemeyer, who is seated in the aluminum-clad streamlined version of the Auto Union race car. (Porsche Museum Archives)

BELOW LEFT: 1895, Ferdinand Porsche (seated at far right) at a meeting within the electric company Béla Egger & Co. in Vienna. A young Porsche was hired in 1893 and within four years had been promoted to manager of testing and assistant in the calculations office. (Porsche Museum Archives)

BELOW RIGHT: 1900, Ferdinand Porsche at the steering wheel of a Lohner-Porsche electric race car. With this automobile Porsche set his first driving record in September 1900, taking fourteen minutes and fifty-two 1/5 seconds to travel ten kilometers with a 400-meter incline, outpacing the fastest gasoline-powered cars. (Porsche Museum Archives)

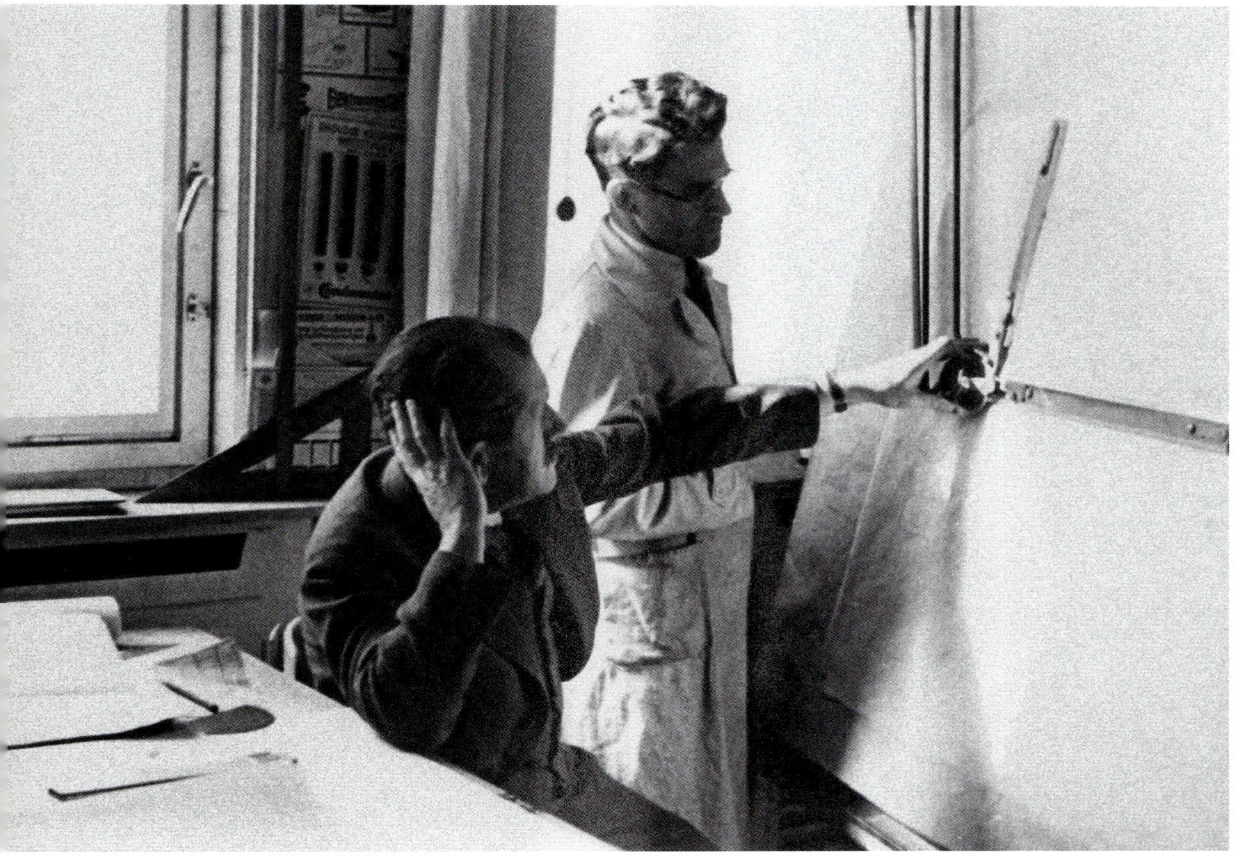

Ferdinand Porsche in the Kronenstrasse design offices, Stuttgart, circa 1937. (Porsche Museum Archives)

Ferdinand Porsche returned to Austria and the technical directorship of Steyr, for which he designed important new models. But Austria's banking institutions were hit hard by the stock market collapse of October 1929. Steyr would be absorbed into a group that already included Austro-Daimler and Puch, another Austrian motor maker. Winding up Porsche's three-year contract after only a year, Steyr placed the engineer at liberty from April 1930, at age fifty-four.

"At my age I shouldn't take a step backward!" Porsche told his nephew, saying that some firms were keen for him to lead engineering but reluctant to give him a management role. As Porsche summed up to his son Ferry, "It makes no sense for me to keep going from one company after another." Ferry put it differently: "My father found that when he signed a contract with a firm, they could live another ten years on his designs, but he couldn't!"

From these musings arose the idea of setting up an independent design office for automotive engineering. With the help of a small team, Porsche could contract with companies to design and build their vehicles. He could concentrate on the engineering he enjoyed while enduring a minimum of the company politics that he detested.

Ferdinand Porsche's new idea depended on the financial support of Adolf Rosenberger, a successful racing driver and wealthy dealer in iron and steel. Porsche's son-in-law Anton Piëch joined as legal advisor while maintaining his Vienna law practice. To help realize his designs, Porsche turned to a lieutenant of old, Karl Rabe, who was lured from Austro-Daimler as chief designer and member of the management board on a four-year contract.

Others attracted to Porsche's new studio in Stuttgart were Steyr's Josef Kales, Josef Zahradnick, and Karl Frölich. After postgraduate tutoring by Steyr's Walter Boxan, Ferry Porsche joined the team, as did Ferdinand Porsche's nephew Ghislaine Kaes, who had been exposed to all the departments at Steyr. Franz Xaver Reimspiess came from Austro-Daimler, as did Otto Zadnik; brilliant theoretician Josef Mickl was a veteran of Porsche's prewar activities.

Steyr and Daimler-Benz were the training grounds for talented body designer–engineer Erwin Komenda, youngest of the new team after Ghislaine and Ferry. Porsche's grandson Ferdinand Piëch remembered Mickl and Komenda as "real biters; they bit right in and performed brilliantly."

The going was hard in Depression-era Europe. In 1931, said Ferry, "there were months when I drew no salary at all, for the simple reason that we had no funds to back us up." He admitted that "we probably had far more technical skill and enthusiasm than business acumen."

"It was bad, very bad," added Ghislaine Kaes. "Sometimes, when the money ran out completely, Herr Rosenberger would give me the keys to his car and tell me to go to a certain address in Pforzheim and collect DM 10,000, which was a lot of money. In 1932 there was not only no money, there were no contracts."

With the inauguration of Germany's Third Reich in 1933, however, and its encouragement of Germany's motorization, Porsche's outlook brightened, and consulting contracts began. The new Auto Union combine needed design services; for its Wanderer marque Porsche created a new model. Auto Union also backed Porsche's design of a new Grand Prix car for 1934 to promote its four brands, symbolized by its four-ring emblem.

This brought to life the magnificent Auto Union Grand Prix car with its V-16 engine behind the driver. Immediately competitive with archrival Mercedes-Benz, in the hands of racing driver Bernd Rosemeyer it was the outstanding Grand Prix car of the 1936 season. The Auto Union cars also set numerous speed records. The contract to design and support the racing cars was a windfall for Porsche, for its terms awarded him a substantial share of the subventions paid by the German government.

The Auto Union racing car's success was a splendid calling card for the Porsche office when Germany's automakers, and later its government, decided to tool up to build the "people's car," the Volkswagen, a publicity triumph of Hitler's regime. The VW's ultimately global success derived from its logically radical design, a product of the genius of Ferdinand Porsche and his hand-picked team.

The Volkswagen project contributed to a rapprochement between Porsche and Daimler-Benz. The Stuttgart automaker was active in building the VW prototypes. It fabricated steel bodies for some of the first five test cars of 1936 and built the next thirty improved cars, which were ready for testing at the end of 1937.

Another 1936 initiative that drew Porsche and Daimler-Benz together was engineered by racing driver Hans Stuck. Keen to bring the world land-speed record to Germany, Stuck had promises of a Mercedes-Benz engine from the Air Ministry and a car design by Porsche. With these assets, in the summer of 1936, he approached Daimler-Benz to see whether it would be willing to build the car. He received an assent that led to the building of a six-wheeled record breaker, Porsche's Type 80. The onset of war kept it from being completed and run.

Porsche and his team also acted as consultants to Mercedes-Benz on its new racing cars for 1938. They advanced novel designs including two-stroke engines and car layouts both front- and rear-engined. Meanwhile they experimented with sports cars, including a design of their own and variants of the VW. These generated exciting ideas that they stored up for future exploitation.

Porsche Type 64 Berlin-Rom racer on the autobahn in Germany. (Porsche Museum Archives)

Moved for safety during the war to a sawmill at Gmünd in rural Austria, the Porsche engineers decided to earn their keep in peacetime by fashioning sports cars from Volkswagens. Though in his seventies, Ferdinand Porsche still had much to contribute. Twin carburetors and aluminum bodies helped his first cars astonish by their fleet performance—still a characteristic of Porsche cars today.

In addition to spending more than twenty years in the motor industry as a consultant and as an executive with Fiat, Ford, and General Motors, Karl Ludvigsen has won numerous awards for his journalism and books about automotive history and racing cars and drivers. He lives in Suffolk, England, with his wife, Annette. His work can be seen at www.karlludvigsen.com.

DAN NEIL

PORSCHE AND THE CONQUEST OF THE AIR

Porsche deserves attention because the company produces an object with special cultural redolence: the 911 Carrera, seventy-five years in the making, remains the defining and essential sports car in what is now Porsche's diversified portfolio. Evolving patiently and iteratively over decades, the 911 is like a telescope through which we see across light years of automotive history the twinkling of an immense explosion.

The late 1930s in Germany was a period of high rationalism in car design. The Reich's interest in aerodynamics and fuel efficiency had a darker purpose, but under its patronage, the domestic auto industry flourished. Germany's "Silver Arrows" dominated Grand Prix motorsport; meanwhile, select groups of technicians inside the carmakers were engaged in scientific propaganda, building radical streamliners such as the Mercedes-Benz T80 and the Auto Union Type C Stromlinienwagen, to set world speed records.

It was into this milieu that Ferdinand Porsche brought the Type 64 Berlin-Rom Rennwagen (1938/9): a long-distance streamliner designed to compete in a proposed Mille Miglia-like road rally between Berlin and Rome, subsequently canceled with the outbreak of hostilities.

If you had seen the Type 64 wooden styling buck (a contoured frame over which the body

Porsche Type 64 Berlin-Rom Racer, 1938, Collection of the Automuseum Prototyp, Hamburg, Germany, Photograph © 2013 Photographers Hamburg

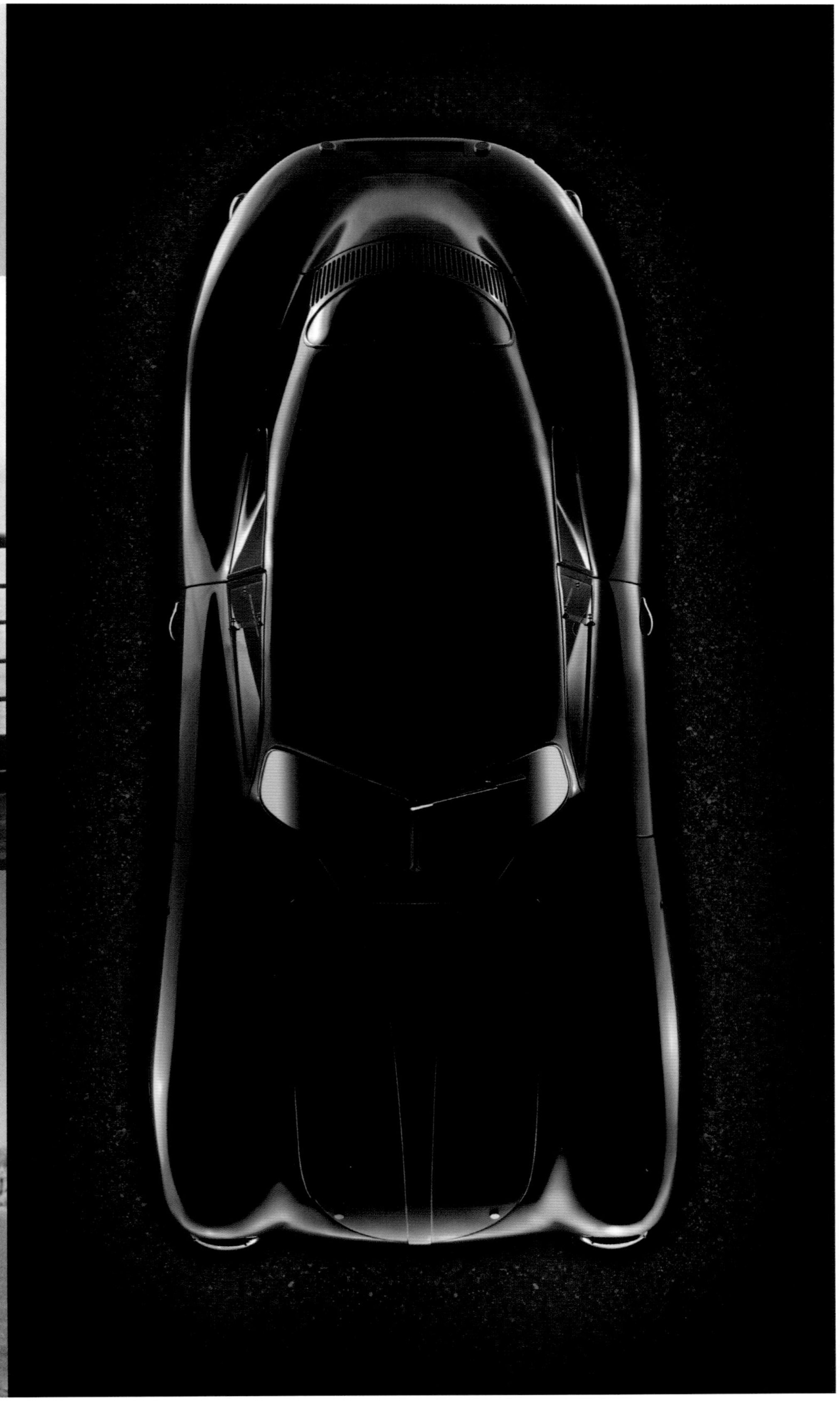

Ferdinand Porsche at age 22 in 1897; arriving in New York during a visit to the U.S. in 1936; and with a Volkswagen Beetle in the mountains of Austria in 1950. (Porsche Museum Archives)

Porsche Type 64 Berlin-Rom Racer, 1938. Collection of the Automuseum Prototyp, Hamburg, Germany. Photograph © 2013 Photographers Hamburg

panels were shaped), you would have admired the car's fluid harmonies in skeletonized form. The car's canopy is a mere blister, designed to minimize drag. The wheel openings are enclosed with fender skirts for the same reason. The car's conformation is dictated by its mechanical starting point, the Volkswagen Type 1, which began as Adolf Hitler's "people's car."

The rear-engine Type 1—distinctive for its rounded carapace and rear louvered engine intake (designed to steal cooling air from the car's laminar flow)—was the National Socialists' biggest gamble on scientific car design. The proof of the concept would have to wait until after World War II and the company's unlikely resurrection in British-occupied Wolfsburg. The VW Beetle would go on to become one of the best-selling cars in history.

It's useful here to note the difference between what the Germans were doing—pursuing, in a programmatic way, applied aerodynamic research in the lab and in the field—and what was happening in other European countries and in the United States, where aerodynamics took a less rational and more decorative turn. From the days of the torpedo-shaped "La Jamais Contente" (1899), designers and engineers understood that the way an automobile moved air aside was a complex problem in fluid dynamics that affected stability, handling, top speed, engine cooling, and fuel efficiency. But in the grand era of automobiles between the wars, "fast" bodywork consisted of teardrop-shaped pontoon fenders and long, lyrical scrolls of sheet metal, geometries that represented mere aerodynamic guesswork.

In the United States, a plausibly aerodynamic car design had been tried—the Chrysler Airflow (1934)—and it was not a commercial success. In Detroit, with few exceptions, strict aerodynamic efficiency would remain subservient to styling for the next fifty years.

The European auto industry continued its research until the eve of World War II, building prototypes with aerodynamic efficiency (measured as a coefficient of drag, the lower the better) that rival the best carmakers can claim today. For example: a Chevrolet Volt electric car has a coefficient of drag of 0.28. The experimental Tatra Schlörwagen (1939, built in Czechoslovakia) claimed a drag coefficient of a mere 0.15.

The presence of the Type 64 in this exhibition nicely recapitulates this era of high rationalism—the applied wisdom of men such as aeronautical engineer Edmund Rumpler; Hans Ledwinka and Paul Jaray (of Tatra); and Porsche's own chief body construction engineer, Erwin Komenda.

As well, the Type 64 represents the creative and technical starting point for Porsche after the company was reconstituted in exile, in an old sawmill in Gmünd, Austria. Three years after V-E Day, during most of which time Ferdinand Porsche himself was unjustly detained in France, Porsche had produced its first 356 "Gmünd" coupe, a car with a radically

Porsche Type 356/2 Gmünd Coupe, 1948, with wool tufts attached in order to research aerodynamics. Photographed at the Porsche factory in Gmünd, Austria. (Porsche Museum Archives)

low 0.30 drag coefficient, better than a modern 911 Carrera. The 356's disciplined approach to aero efficiency has inflected Porsche design ever since.

Finally, the Type 64 helps us measure the depth of historical obligation, the respect for precedent embodied in the modern 911 Carrera. The framing of both cars is the same: compact, broad, tautly form-fitted, with a deep set on its wheels, a shape comprising two essential lines—the fender line and roof line—converging at a pinched tail. Yes, the dimensions of the cabin greenhouse are different, but the two designs are ineffably and unmistakably by the same hand.

Why, after seventy-five years and several revolutions in automotive technology, should the 911 Carrera bear any resemblance to the Ur-Porsche of 1939, or subsequent models in 1948, or 1963, or 1997? What is served by this fealty? No other automotive company—not Mercedes, not Cadillac—is obliged to worship its ancestors in quite this way. And yet such is Porsche's belief in the 911 that the company hands off, generationally, plans for the next optimum sports car that certainly aren't optimum from a technical perspective.

The technical term for a design artifact is a

The framing of both cars is the same: compact, broad, tautly form-fitted, with a deep set on its wheels,

skeuomorph, a decorative or emotionally significant element that recalls an obsolete technology, retained (deliberately or not) in the interests of comfort and familiarity: the watch pocket in Levi's jeans, for example, or the tachymeter bezel on a mechanical watch.

The 911 Carrera, in ways large and small, is such an artifact. The shape of the 911—its basic ratios, its balance of volumes, its phenotype—comes to us directly as an iteration of the Type 64 and the era of scientific design. As horsepower and top speed have increased over the decades, Porsche's researchers have invested thousands of man-hours neutralizing the potential instability inherent in the coupe's rear-engine design (the company's history of rear spoilers is a saga unto itself). As the years roll on, the 911 Carrera has evolved stylistically, aerodynamically, and functionally, but its basic appearance, beloved by enthusiasts, is immediately recognizable. At the same time, with the Boxster, the Cayman, and the latest 918RSR, Porsche offers the latest advances in mid-engine technology.

The defining irony of Porsche is that for the 911, its most revered model, this proudly technical, fiercely logical company has enthroned what could be considered obsolete technologies for what are essentially emotional reasons. That makes the cars an expression of belief, a *Kulturwerk*, and therefore a legitimate object of this museum's curatorial interest.

Dan Neil is the automotive critic for the Wall Street Journal *and winner of the Pulitzer Prize for Criticism.*

Technical Notes

Engine Type: 1,131 cc, air cooled, horizontally opposed, 4-cylinder
Horsepower: estimated 50 hp at 4,000 rpm
Transmission: 4-speed manual
Top Speed: 95 mph
Wheelbase: 94.5 in.
Suspension: Independent front transverse torsion bars with parallel trailing arms; independent rear swing axle with a single trailing arm on each side
Weight: 1,287 pounds

PORSCHE TYPE 64 BERLIN-ROM RACER

This trendsetting coupe is truly Porsche's "ancestor car." Its frame design, rear-engine location, aerodynamics, and shape anticipated all postwar Porsche cars. In 1931 Dr. Ing. Ferdinand Porsche, founder of the dynasty, opened a design consulting office in Stuttgart, Germany. A brilliant engineer and theoretician, Dr. Porsche was responsible for the original Volkswagen. In 1937 he and his staff, including his son Ferdinand ("Ferry"), began work on a sports car version of the VW, called the Type 64. As his firm became successful, Porsche decided to build his own sports car, to be called the Type 114 or F-Wagen. The mechanical drawings were never completed, but by January 1939 the car had been conceptualized in detail.

The Type 114's engine was to be located between the passengers and the rear axle, with the transmission placed abaft the axle. Porsche planned a then-radical 1.5-liter V-10 power plant, anticipating present-day Formula One engines. Designer Erwin Komenda styled the curvaceous and startlingly modern body, which resembled an aerodynamic VW coupe. Porsche anticipated producing this car as soon as world tensions subsided.

At that time planning began for a major Berlin-to-Rome race. Despite the start of war, major automakers created special cars for the event. Porsche and his colleagues convinced the National Sporting Authority (ONS) to commission three sports coupes based on the VW Type 60 (a.k.a. VW Type 60K10 Rekordwagen). Komenda's design had fully skirted fenders and a narrow cabin; the seats were staggered to permit room for two occupants. Twin spare tires were located under the hood, as was a large fuel tank. The new car was ready, but the ill-timed race was canceled.

The three coupes were then fitted with horns and driven on the road. One was damaged in a collision; Ferdinand Porsche and his chauffeur used the other two.

Only one complete Type 64, formerly owned by the Porsche family, survived the war and is now in private hands. This Type 64 re-creation was assembled on a surviving Porsche-built VW chassis with many original parts from the number 2 Type 64. The new alloy body took three years to build.

a shape comprising two essential lines—the fender line and roof line—converging at a pinched tail.

Porsche Type 64 Berlin-Rom Racer, 1938, Collection of the Automuseum Prototyp, Hamburg, Germany, Photograph © 2013 Photographers Hamburg

CAM INGRAM RESTORATION NOTES

"In the nascent days of car collecting, the established Concours d'Élégance events, such as Pebble Beach, praised cars that were restored to original spec. Many restorers believed that the only way to impress the judges and achieve a 'perfect' 100-point score was to make the car look new again. Within the last decade or two, however, ideologies began to change. Perhaps the primary idealistic shift in the car-collecting mentality of recent times is that preservation is now seen as being more desirable than restoration. Today, an identical model in a preserved, original state will often bring more money than a 100-point Concours-restored example."

DAVID SWIG, BONHAMS AND BUTTERFIELDS AUCTIONEERS CORPORATION

David Swig's statement not only reflects the paradigm shift in car collecting but also illustrates the change in automotive restoration philosophy. "They are only original once" has become the new mantra for the collector car market. Yet, what are the new criteria for restoration when a car cannot be saved or preserved?

An authentic restoration not only restores function and appearance but also honors the history of the car and the people who built it. In other words, the car should be treated as a historical artifact and not re-created in our idealized notion of perfection. Modern restoration philosophy puts the emphasis on material authenticity instead of the temptation to over-restore. The true craft of automotive restoration seeks to make the restoration invisible.

An authentic restoration thus begins with research and documentation. Whether it's traveling to look at original examples, devoting countless hours to scanning period photographs, or reading obscure resources, undertaking credible research first is the only way to authentically restore a car.

This attention to detail goes well beyond the type of paint and panel fit of any restoration project. It addresses the type of fasteners used during a period, for example, and the specific construction methods used in the various automobile factories around the world. A thorough restoration addresses every minute detail, from the correct windshield wiper motor to the correct grain in the leather interior hide that was used circa the period of the car. It's only with this completed research that one starts the slow disassembly process of a historically significant automobile, documenting every step with digital photography.

All these principles were kept in mind when Road Scholars Restorations undertook the restoration of the 1950 Gmünd coupe belonging to Hans-Peter Porsche. His 1950 Gmünd was the forty-fifth of the first fifty-two Porsche cars built in a converted sawmill in Gmünd, Austria, after World War II (the reason the first fifty-two Porsche cars are referred to as Gmünds).

Road Scholars spent two thousand hours on international travel and Gmünd research. Eight different Gmünds were inspected, and two were laser scanned for body data points. These data points allowed Road Scholars to create the exterior body buck and replicate the hand building process of the car.

Once 045's original body was stripped bare, it was clear that the ravages of time and previous paint jobs had ruined many of its body panels—whereupon Road Scholars coachbuilder Hans Sahling used his power hammer to replicate the needed panels and gas welded them in place just as the factory had done. The hammer marks from mating the body to the frame were left undisguised, just as in the original, and the uneven rivet pattern was also replicated. To impede electrolysis between the steel chassis and aluminum body, a simple strip of herringbone cotton, exactly replicated from the original material used in 1950, was mounted between the two surfaces to separate them.

A hint of original red paint in the front trunk remained and was painstakingly matched after many attempts. Eventually 045 would be

ABOVE: Two views of a Porsche Continental body shell in Road Scholars Restorations workshops. Photographs © 2013 Cam Ingram. **MIDDLE: Porsche Type 356/2 Gmünd Coupe,** 1948, production in Gmünd, Austria. At left is Porsche's longtime chief body construction engineer, Erwin Komenda, with supervisor Otto Husslein. (Porsche Museum Archives) **BELOW: Hans-Peter Porsche's** award-winning Gmünd 045 at Pebble Beach in 2011. Photograph © 2013 Cam Ingram.

painted in single-stage paint as it originally had been in 1950. Paint drips were left unsanded in the engine compartment bay, as was documented on an original Gmünd. Further maintaining authenticity, Road Scholars applied the undercoating with a paintbrush.

Countless hours were spent repairing the original engine's degraded magnesium cooling fins. This original engine was actually from a Kübelwagen, the German military jeep during the war. With the lack of resources after the war, supply was wholly unpredictable. Porsche KG would use at least four different engines in its aluminum 356s (Gmünd cars), sourcing them from KdF Beetles, Kübelwagens, Schwimmwagens, and industrial applications.

Parts manager Brandon Terretti searched the world over for the obscure SWF-brand "frog's-eye" taillight lenses. These would eventually be found on a shelf in the Netherlands in original condition.

After a marathon search for a rear license plate light, Road Scholars was fortunate enough to have Jerry Seinfeld lend us the light off his original Gmünd coupe. Upon receipt machinist Bill Woodruff created a copy that is indistinguishable from the original but for patina.

Finally, an original broken early Porsche script was sourced for the nose of the car. The Road Scholars team sand cast it and smelted World War II German canteens to replicate the quality of the aluminum. Thus the finished script had the same flawed casting details and porous aluminum look as the original.

All these authentic details and dedication to the restoration process "to make like original" were recognized and awarded on the lawn at Pebble Beach in 2011. Hans-Peter Porsche took first place in Post War Sports Cars Closed with the Gmünd coupe. It was the first time a Porsche had won such a distinction at Pebble.

As the study of the automobile and car collecting have entered a new era of appreciation, it's clear that automobile restoration has evolved as well, with the auto restorer's job now to conserve each project as a historical and cultural artifact. Many works of art and artifacts are precious to human history, and many people desire to keep them for future generations. Preserving those pieces can be tricky because the art and artifacts are constantly facing damaging influences, such as rot, deterioration, and faulty conservation or restoration methods.

With this new approach to restoration in the collector car world, future generations will be able to view cars in their original intent.

Cam Ingram is the co-owner of Road Scholars LLC, a company that specializes in restoring significant automobiles and managing car collections.

RIGHT: Porsche Type 356 Gmünd Coupe, 1949, Courtesy of the Ingram Collection, Photograph © 2013 Michael Furman

KEN GROSS

WHY PORSCHE?

Automobile exhibitions in fine art museums are hardly unprecedented. In autumn 1951 the Museum of Modern Art in New York City presented *Eight Automobiles*, describing it as "an exhibition concerned with the aesthetics of motorcar design."

Arthur Drexler, then curator of architecture and design for MoMA and curator of the exhibition, wrote: "Automobiles are hollow, rolling sculpture. They have interior spaces corresponding to an outer form, like buildings, but the designer's aesthetic purpose is to enclose the functioning parts of the automobile, as well as its passengers, in a package suggesting directed movement along the ground. The illusion of movement is inappropriate to architecture, but it is often applied to sculpture." —(*EIGHT AUTOMOBILES*, A CATALOGUE FROM THE MUSEUM OF MODERN ART, NEW YORK, 1951)

LEFT: Ferry Porsche in 1958 at the Porsche assembly line in Stuttgart-Zuffenhausen. (Porsche Museum Archives)

ABOVE: An October 1950 Porsche brochure illustrates the early technological breakthroughs of company founder Ferdinand Porsche that led to the Porsche Type 356. (Porsche Museum Archives)

RIGHT: Erwin Komenda, "Ferry" Porsche, and Ferdinand Porsche with the first Volkswagen-based "special" roadster built in Gmünd, Austria. (Porsche Museum Archives)

Considerable time elapsed before other major fine art institutions followed suit. But in the past ten years, themed automotive exhibitions have appeared at the Museum of Fine Arts, Boston; the Phoenix Art Museum; Les Arts Décoratifs at the Louvre in Paris; the High Museum of Art (Atlanta); the Portland Art Museum; the Utah Museum of Fine Arts (Salt Lake City); and the Frist Center for the Visual Arts (Nashville).

And now it's North Carolina's turn, with a somewhat different approach: *Porsche by Design: Seducing Speed.*

The history of the automobile is replete with notable father-and-son pairings: Ettore Bugatti and his son Jean; Henry and Edsel Ford; Enzo Ferrari and Alfredo, better known as Dino; Renzo Rivolta and Piero, to name a few. But not one of these duos together achieved the enduring success of Dr. Ing. h.c. Ferdinand Porsche and his son Ferry. Moreover, Porsche family accomplishments didn't stop with the work of the founding father and his son. Ferry's sons Ferdinand Alexander (better known as Butzi) and Wolfgang, as well as Dr. Ferdinand Porsche's son-in-law Dr. Anton Piëch and Piëch's son Ferdinand (the present-day chairman of the Volkswagen Group, which owns Porsche AG) have all made major contributions.

Fine art museums in Montreal, Melbourne, and Tokyo have celebrated the work of the Bugatti family, but until now there has never been a fine art museum exhibition devoted solely to Porsche. *Porsche by Design: Seducing Speed* presents the automobile in an artistic and design context by displaying the production of a single German company and its founders, the remarkable Porsche family.

Porsche cars have had a long and storied history. Dr. Ing. h.c. Ferdinand Porsche developed the world's first hybrid car in 1897; he later worked at Austro-Daimler, designed and developed racing cars for Mercedes and Auto Union, and led a successful design consultancy in Stuttgart, Germany, which resulted in the Volkswagen. The Porsche formula for a lightweight, aerodynamic, and sporty car, with a sturdy platform frame, powered by an air-cooled, horizontally opposed, rear-mounted flat four-cylinder engine, began in 1938/9 with Porsche's VW-based Berlin-Rom Type 64 racing car.

In 1948 Dr. Porsche's son Ferry (also Dr. Ferdinand Porsche) built a Volkswagen-based "special" roadster in Gmünd, Austria, that received international acclaim after a Swiss magazine published a favorable road test. The first Porsche series production car, a coupe designed by Erwin Komenda, was known as the 356, because it was job number 356 in the design studio's production. Sleek, compact, streamlined, and innovative, the tiny two-seater used a strong platform frame and a modified VW 1,131 cc air-cooled four-cylinder engine, mounted behind the rear axle. Its suspension was independent all around with torsion bars, front trailing links, and rear swing axles. Despite its cable brakes and 85 mph top speed limitation, it handled extremely well. Production Porsche 356s were soon successfully campaigned in racing and rallying by a growing cadre of enthusiastic owners.

Porsche Type 356 Gmünd Coupe, 1949, Courtesy of the Ingram Collection, Photograph © 2013 Michael Furman

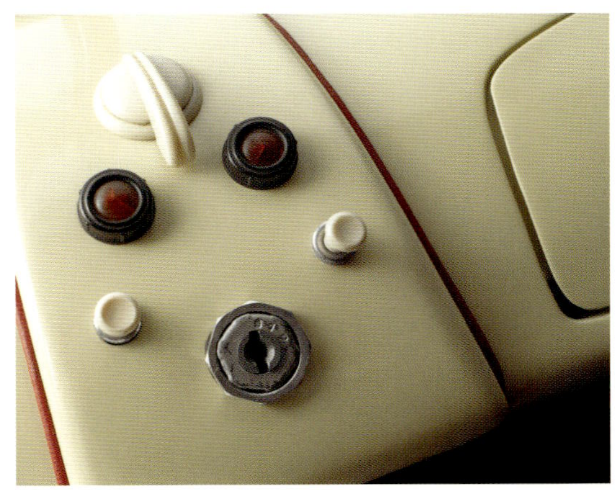

Porsche Type 356 Gmünd Coupe, 1949, Courtesy of the Ingram Collection, Photographs © 2013 Michael Furman

The Porsche Design Studios returned to Germany in 1950. Car production increased, with the bodies initially made by Reutters. Porsche's financial success was supplemented by a royalty on each Volkswagen built, and from licensing a clever Porsche synchromesh transmission system initially developed by Dr. Porsche for an ultimately unsuccessful Cisitalia Grand Prix car. Porsche owned Allgaier, a company that built a range of versatile diesel-powered farm tractors. The elder Porsche died early in 1951, but not before he saw his cars beginning to challenge the world's best.

Lacking large advertising budgets, Ferry Porsche understood that racing was the crucible in which to prove Porsche's superiority and capture world attention. A single Porsche qualified for the 1,100 cc class at the Le Mans 24-Hour race in 1951, won in its first appearance there, and finished twentieth overall. Enthusiastic GIs brought a few Porsche cars home from Germany. On these shores, imported by Max Hoffman, an Austrian expatriate who was a friend of the Porsche family, the low, streamlined little cars soon attracted considerable interest and a rabid following. The one thousandth Porsche was sold in 1951. Porsche would repeat its Le Mans class-winning feat the following year.

Ferry Porsche knew that modified production cars were not sufficient to win major races. Earlier, modified privateer Gmünd coupes, a series of fifty alloy-bodied cars built by Porsche in Austria, and the Glöckler-Porsche racing cars (built by Frankfurt VW distributor Walter Glöckler and his workshop chief, Hermann Ramelow) proved Porsche design principles could be very competitive in small-displacement classes. Project 550, a purpose-built racing coupe designed for the lowest possible aerodynamic drag, began in 1952.

In the early 1950s, Porsche victories at Le Mans, in Mexico's La Carrera Panamericana, and in countless other venues established the small Stuttgart, Germany–based automaker as a "giant killer." Increasingly sophisticated Porsche racing cars challenged and beat the world's best, and Porsche production cars developed a loyal following worldwide. The 356 models morphed into the wildly successful 911s.

Porsche always took competition seriously. The company marshaled its best engineers, led by Dr. Ferdinand Piëch, to ensure its cars would be as advanced as any in the world. Legendary race cars like Porsche's vaunted, technically sophisticated, all-conquering 917s were virtually unmatched in their era by any competitor, including Ferrari and Mercedes-Benz. Today Porsche is a full-line manufacturer with a wide range of 911, Cayman, and Boxster sports cars, Cayenne sport utility vehicles, and the Panamera four-door sedan. Six decades of popularity in America have established Porsche firmly on the top tier of performance automakers.

Today a modern Porsche 911 evokes the shape and spirit of the earliest Type 64. The studied continuity of Porsche evolution is part

The studied continuity of Porsche evolution is part of the marque's fascination. The arch of the roofline, the taut stance of the wheels, the car's sense of purpose and motion are all unmistakably Porsche.

Technical Notes

Engine Type: 1,098 cc, air cooled, horizontally opposed, OHV 4-cylinder
Horsepower: 72 hp at 5,000 rpm
Transmission: 4-speed nonsynchromesh manual
Top Speed: 100 mph
Wheelbase: 82.7 in
Suspension: Independent front transverse torsion bars with parallel trailing arms; independent rear swing axle with a single trailing arm on each side
Weight: 1,430 pounds

Porsche Type 356/2 Gmünd Coupe, 1948, on location in Gmünd/Kärnten, Austria. (Porsche Museum Archives)

of the marque's fascination. The arch of the roofline, the taut stance of the wheels, the car's sense of purpose and motion are all unmistakably Porsche. A Porsche's ignition switch is still located to the left of the steering wheel, as it was designed years ago for a speedy Le Mans start. When Porsche changed from air cooling to water cooling in the 1990s, engineers ensured the characteristic steely whir of the cooling fan remained largely the same. Porsche, as a car company, is all about respecting tradition while constantly improving. While rivals have changed radically though the years, and not always for the better, Porsche cars have evolved steadily, using the same principles of design, streamlining, engine configuration, sophisticated engineering, and sporting ability they have had since the company's inception.

No doubt Dr. Porsche would be very pleased.

An automotive journalist for forty years, Ken Gross is former director of the Petersen Automotive Museum and a Pebble Beach Concours judge for twenty-five years. He consults for several museums and is guest curator of Porsche by Design: Seducing Speed *at the NCMA.*

Porsche Type 356 Gmünd Coupe, 1949, Courtesy of the Ingram Collection, Photograph © 2013 Michael Furman

PORSCHE TYPE 356 GMÜND COUPE

This rare early Porsche, 356/2-017, one of the oldest known Gmünd coupes extant, is the seventeenth example of just fifty to fifty-two built.

Body construction manager Erwin Komenda's lovely teardrop-shaped design for the first Porsche production coupes set the style for all Porsche cars to come. To escape Allied bombing, the company had relocated from Stuttgart, Germany, to a sawmill in Gmünd, Austria. The model made there, known as project 356, from a sequence that started in 1931 when Dr. Ferdinand Porsche opened his independent design studio, was overseen by his son Ferry.

Gmünd coupe bodies were painstakingly hand-hammered from sheet aluminum over wooden bucks. The chassis was a sturdy pressed-steel platform, formed in unit with the floor pan. Fully independent suspension all around, excellent aerodynamics, inherent rigidity, and light weight made the Gmünd a natural race car.

Ferry Porsche astutely realized that, lacking a large advertising budget, Porsche would have to capture world attention through racing. In September 1951 a modified Gmünd coupe with an experimental 1,496 cc engine averaged 94.66 mph for seventy-two hours on the Montlhéry banking in France, setting a remarkable endurance record for a car in that class. Encrusted with bugs and oil, it was proudly displayed at the Paris Salon.

Orders started coming, and in late 1949 the Reutter Company in Stuttgart was given the assignment to construct 500 steel-bodied coupes. In 1950 Porsche moved back to Stuttgart, where more space and access to materials allowed for faster production. By year's end Porsche had sold 298 units.

Some cars were converted to SL (*Sport Leicht*) racing specification. Gmünd coupes raced successfully at the Le Mans 24-hour race as well as in rallies, notably Liège-Rome-Liège, where one finished first in class and third overall. When race team leader John Wyer first saw the new Porsche cars at the Sarthe circuit, he quipped, "Interesting...but the engine is in the wrong end."

By the time Porsche stopped producing the 356 in 1965, the company had made 76,000, of which the Gmünd coupes remain the most highly prized.

CALIFORNIA
17TH

MAY
CALIFORNIA
XEL 019
2013
PORSCHE
1600 SUPER

Porsche Type 356 Speedster 1600 Super, 1958, Collection of Chad McQueen, Photograph © 2013 Peter Harholdt

Previous owner Steve McQueen (left) sold this Speedster to collector Bruce Meyer (right) and then bought it back in 1974. (Photo courtesy Bruce Meyer)

In 1955 the Porsche Type 356 Speedster was advertised at just under $3,000 in the United States, available at the New York City dealership owned by Max Hoffman. Also at left, a glimpse inside the 1950s Park Avenue Hoffman dealership designed by Frank Lloyd Wright, with a white Speedster at center stage on a rotating platform. (Porsche Museum Archives)

PORSCHE TYPE 356 SPEEDSTER 1600 SUPER

Porsche's beloved Speedster began its life as a "stripper." U.S. distributor/entrepreneur Max Hoffman convinced his old friend Ferry Porsche that there was a market for a bare-bones roadster that could easily be converted for racing. Priced at just $2,995 (the heater and tachometer were extra), the first Speedster was a lightweight version of the 356 Cabriolet. The Speedster's distinguishing feature was its low windscreen, chopped 3.5 inches shorter than the Cabrio's. With its canvas top erected, the Speedster resembled a short, fattish man with his hat pulled over his eyes. Speedsters were quickly nicknamed "bathtubs."

Under the early Speedster's single-vent deck lid was a 64 hp, 1488 cc "Normal" flat-four engine. For $500 more you got the hotter 82 hp "Super" engine. Speedster gearboxes had lower first and second gear ratios, and the cars accelerated faster than the heavier coupes. In 1956 wraparound teardrop taillights replaced the "beehive" domes. By 1957 three versions were offered: the 70 hp Normal, the 88 hp Super, and the race-ready 128 hp Carrera. Properly geared, a Carrera could top 125 mph. Actor James Dean briefly raced a 356A Speedster before he bought his ill-fated 550 Spyder.

In 1959 Porsche dropped the Speedster in favor of the Convertible D. Just 3,122 Speedsters were sold, making them rare and valuable today. At any Porsche get-together, Speedster drivers are accorded high honors—none more so than the previous owner of this car: "the king of cool," actor, racer, and all-around tough guy Steve McQueen. McQueen purchased this car in 1958, raced it in Sports Car Club of America competition, and drove it for years. He sold it to collector Bruce Meyer and then bought it back in 1974. It doesn't have the Speedster and Reutter emblems, because they were removed when McQueen owned the car, and his son Chad prefers leaving it that way.

Technical Notes

Engine Type: 1,582 cc, air cooled, horizontally opposed, OHV 4-cylinder	
Horsepower: 88 hp at 5,500 rpm	
Transmission: 4-speed synchromesh manual	
Top Speed: 110 mph	
Wheelbase: 82.7 in.	
Suspension: Independent front transverse torsion bars with parallel trailing arms; independent rear swing axle with a single trailing arm on each side	
Weight: 2,150 pounds	

Porsche Type 356 Speedster 1600 Super, 1958, Collection of Chad McQueen, Photographs © 2013 Peter Harholdt

Porsche Type 356B 1600 Carrera GTL Abarth Coupe, 1961, Courtesy of the Ingram Collection, Photograph © 2013 Michael Furman

Porsche Type 356B 1600 Carrera GTL Abarth Coupe, 1961, Courtesy of the Ingram Collection, Photographs © 2013 Michael Furman

Porsche Type 356B 1600 Carrera GTL Abarth Coupe

Every car that designer Carlo Abarth touched went faster. Ferry Porsche and Abarth first collaborated in 1946 on the ultimately unsuccessful rear-engine Cisitalia Grand Prix car that Porsche AG built for Piero Dusio. Abarth had moved to Italy while that project was ongoing. He remained there after it was done, working his magic on special Fiats and Alfa Romeos. FIA competition rules permitted a special body on a production car chassis, just as long as its weight exceeded 1,800 pounds. Porsche contracted Abarth to build a small, high-performance, *über*-lightweight racing series.

Abarth worked with several Italian subcontractors; Franco Scaglione styled this sleek alloy-bodied competition coupe on a narrowed Porsche 356B Carrera 2 floorpan. All twenty or twenty-one examples (sources disagree) were handcrafted in Milan by Carrozzeria Zagato, specialists in building lightweight aluminum bodies.

Viewed alongside a conventional steel 356B, the Abarth is slimmer, smaller, more athletic, and less constrained by the classic Porsche roundness. With a redesigned lower roofline, the Abarth had a low extended nose and was more aerodynamic than the production 356B. To better cool the rear-mounted engine, there's an adjustable rear air scoop and nearly fifty louvers.

In its first iteration, the model 692/3 four-cam engine produced 132 hp; with the succeeding 693/3A, output was upped to 147 hp, and in the last version, power increased to 155 hp. As befitting a racing machine, there's a Spartan cockpit with an Abarth-specific instrument panel fashioned from a sheet of aluminum. The deep bucket seats resemble those of a Porsche Speedster but with better side bolsters.

The first GTL was finished in the spring of 1960. Porsche entered a new Abarth in the grueling Sicilian Targa Florio; it won its class and finished sixth overall, against much larger cars. Entered at Le Mans the same year, the Porsche factory Abarth Carrera won its class and placed eleventh overall. Abarth Carreras were 1600 cc class winners in 1962 at Sebring and the Nürburgring.

This car, number 1008, the eighth Abarth Carrera GTL built, has the first 693/3A engine. It was purchased by Swedish radio personality Carl-Gunnar Hammarlund, who was also an amateur racing and rally driver. Hammarlund's mechanic added weight to the front end to improve the steering response. Hammarlund won two Swedish Grand Touring Championships in this car, scoring eleven wins in eleven starts, after which he retired. The Abarth GTL was sold to a collector in Mexico before the Ingram Collection acquired it. Perfectly maintained, retaining all its original panels and engine, and with just under 11,000 miles, it's a perfect example of one of the rarest Porsche competition cars.

Porsche Type 356B 1600 Carrera GTL Abarth Coupe, 1961, Courtesy of the Ingram Collection, Photograph © 2013 Michael Furman

Technical Notes

Engine Type: 1,587 cc, air cooled, horizontally opposed, DOHC 4-cylinder
Horsepower: 147 hp at 7,400 rpm
Transmission: 4-speed synchromesh manual
Top Speed: 144 mph
Wheelbase: 82.7 in.
Suspension: Independent front transverse torsion bars with parallel trailing arms; independent rear swing axle with a single trailing arm on each side
Weight: 1,762 pounds

A 1960 Porsche poster by Stuttgart-based painter and designer Hanns Lohrer (1912–1995), who captured the Porsche brand in many print materials for the manufacturer during the 1950s and 1960s. Poodles were a recurring theme, as were the bold palette and playfully optimistic sensibility captured in this charming illustration. (Porsche Museum Archives)

LEFT AND RIGHT: Porsche Type 356C Carrera 2 Coupe, 1964, Courtesy of the Ingram Collection, Photographs © 2013 Michael Furman

LEFT AND RIGHT: Porsche Type 356C Carrera 2 Coupe, 1964, Courtesy of the Ingram Collection, Photographs © 2013 Michael Furman

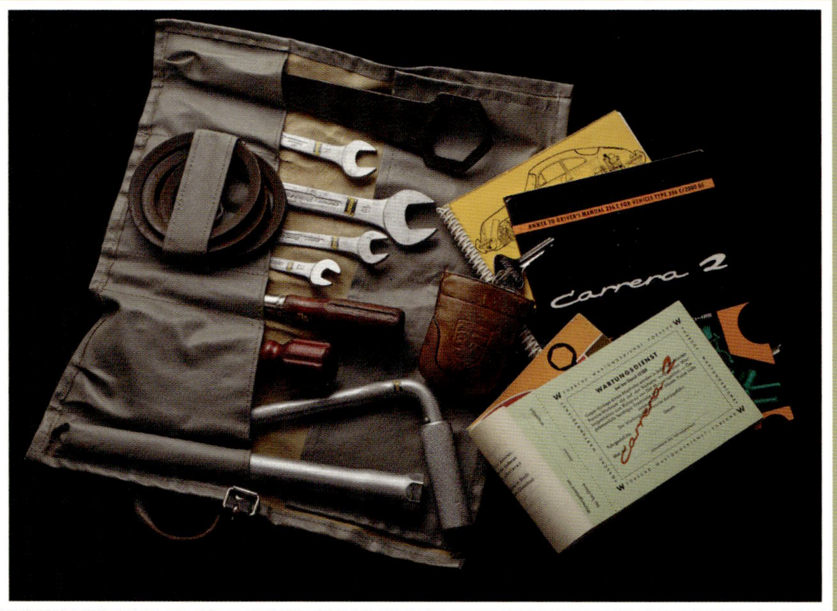

The 356C/SC/Carrera was the ultimate representation of the 356 Series. Taut, compact, nimble, and sporty, the 356 boasted a rugged unit body, torsion bar suspension, a remarkable Porsche synchromesh transmission design, and that now-famous air-cooled, horizontally opposed flat-four engine that would rev willingly, in pushrod form, to 5,800 rpm.

Porsche Type 356C Carrera 2 Coupe, 1964, Courtesy of the Ingram Collection, Photographs © 2013 Michael Furman

Technical Notes

Engine Type: 1,966 cc, air cooled, horizontally opposed, OHV 4-cylinder

Horsepower: 128 hp at 6,200 rpm

Transmission: 4-speed synchromesh manual

Top Speed: 125 mph

Wheelbase: 82.7 in.

Suspension: Independent front transverse torsion bars with parallel trailing arms; independent rear swing axle with a single trailing arm on each side

Weight: 2,150 pounds

PORSCHE TYPE 356C CARRERA 2 COUPE

After fifteen years and considerable success, Porsche's stellar 356 Series, the agile little cars that put the company on the map, were beginning to show their age. While the factory worked in secret on the 356's successor, the last of the 356s were the best.

The 356C/SC/Carrera was the ultimate representation of the 356 Series. Taut, compact, nimble, and sporty, this car was a relative performance bargain. Despite its relatively small size, the 356 boasted a rugged unit body, torsion bar suspension, a remarkable Porsche synchromesh transmission design, and that now-famous air-cooled, horizontally opposed flat-four engine that would rev willingly, in pushrod form, to 5,800 rpm (and in the four-cam iteration, to 6,200 rpm and a bit beyond).

Ken W. Purdy, the leading American auto writer of the day, wrote: "People who own Porsches may be anything from industrial tycoons to freelance artists, but they all have two things in common: (1) When they want to go, they want to go; they don't want temperament, delay, or fussiness in their automobiles; (2) They don't live in ugly houses, they don't wear ugly clothing, and they won't drive ugly automobiles." He concluded, "You almost never see a Porsche driver who looks bored."

Porsche's 356 Series continued to be exported to the U.S. for the 1965 production year. This rare C2 has an optional sunroof, chrome wheels, and twin headrests, making it a very desirable car today that reflects its $350,000 price tag. That's not a bad return on an $8,000 investment for the fastest Porsche road car of its era.

Essayist Jeff Zwart in the passenger seat with driver Tanner Foust during the filming of Zwart's "Family Tree" commercial. Noted collector Jerry Seinfeld is driving the Porsche directly behind Zwart. Photograph © 2013 Jeff Zwart

JEFF ZWART PORSCHE THROUGH THE LENS

High above an abandoned runway, I ask my helicopter pilot to turn and drop altitude. Below me are forty-eight cars that represent the history of Porsche both on the road and in racing. The lens of my remote camera finds the little 1949 Gmünd coupe leading the pack. As we rush toward the collection of Porsches at over 100 mph, the lead car sets a line that takes this history under the helicopter. I count the cars sweeping past the lens and look for the key ones we want to focus on. As the Gulf 917 comes into view, a question instantly hits me. How can I be so fortunate to be not just a part of this moment, but leading this commercial shoot of the launch of the new Panamera?

How did my childhood influence me to get to this point? What drew me to this amazing Porsche brand? It really goes back to a couple of events in my life. My parents bought their first and only new Porsche, a 1964 356C, and

immediately drove me across the country to witness the largest spectator event in the world, the Indianapolis 500. For a nine-year-old, that was big. I witnessed racing in its biggest venue, and our "enabler" to do that was a Porsche. I fell in love with racing and Porsche cars simultaneously.

The other event was going to see the movie *Le Mans* starring Steve McQueen. In the darkened theater, that movie put me in the driver's seat. It made me feel like I was really experiencing what it would be like to race a Porsche 917 down the Mulsanne Straight at top speed. This style of filmmaking attracted me to the craft itself and further embedded my future in the Porsche world.

Today I direct "high-action" television commercials, which has led me to shooting the majority of the Porsche commercials over the last twenty years. It's been a great way to combine my passions and at the same time use my racing intuition. Film was a great outlet for me, since my beginnings were in still photography. Film allowed me to capture the sound and the iconic shape of the 911 in motion, not just still.

The shape of the 911 was dictated by its configuration of having the engine in the rear. This allowed the design to be unique and at the same time timeless. Today, fifty years later, the essential shape remains, and the refinements are spectacular. I am drawn to the shape and the way the light flows across the car's body. From every angle it communicates something different, yet it never separates itself from having a sense of purpose.

Porsche has been a part of American pop culture and seemed to help individuals define themselves by what they drove. Janis Joplin saw fit to not just drive a 356 but to give it a paint job that reflected the psychedelic world she lived in. Paul Newman loved to race and on many occasions found himself behind the wheel of a Porsche, most notably finishing second at the 24 Hours of Le Mans in 1979 in a 935. The two films that launched Tom Cruise's career both had Porsche cars, a 928 and a Speedster. Porsche means something; it represents something; it doesn't simply blend in.

Motorsport has been the backbone of the Porsche brand. One event has been the company's major focus, and that is the most famous

Steve McQueen (right) and his wife, Neile Adams, (in a Porsche Type 911), at Le Mans in 1970. (Porsche Museum Archives)

Steve McQueen in racing gear. (Porsche Museum Archives)

Porsche Type 917 KH Coupe during filming of *Le Mans* with Steve McQueen in 1970. (Porsche Museum Archives)

Janis Joplin atop her Type 356C Cabriolet. The car is featured in the *Porsche by Design* exhibition. RB/Redferns/Getty Images

Porsche Type 356C Cabriolet, 1965, Collection of the Joplin Family, Courtesy of the Rock and Roll Hall of Fame and Museum, Cleveland, Ohio, Photographs © 2013 Peter Harholdt

From the earliest Gmünd coupe to the GT3 R hybrid race car prototype, these are silhouettes developed by the wind.

endurance race in the world, the 24 Hours of Le Mans. Winning this event sixteen times overall, with countless class wins, Porsche has had phenomenal success there, and its pursuit of that success has pushed the design and engineering side of the marque. Shapes and technologies have been developed on the racetrack and transferred to the road cars. Even with all the diverse shapes, there is a familiarity. When you walk through an exhibition like this, you notice how much aerodynamics has played into Porsche's design. From the earliest Gmünd coupe to the GT3 R hybrid race car prototype, these are silhouettes developed by the wind.

My own history in Porsche motorsport has been racing 911s in the U.S. Pro Rally Championship and the famed Pikes Peak International Hill Climb, twelve times at Pikes Peak in nine different Porsche cars. I even had the challenge of competing for Porsche in the Trans-Siberia event that raced us from Moscow to Ulan Bator, Mongolia, in a factory-built racing Cayenne. These Porsche racing experiences have developed my own perspective in filmmaking. The impression that the movie *Le Mans* made on me long ago has now been enhanced by my own racing experiences. I hope my film work has that feeling of pushing something to the ultimate edge with the confidence of its design.

Shooting the Porsche Panamera "Family Tree" commercial allowed me to choreograph the history of Porsche in a unique way. It gave me an incredible look into the cars, not just in a static way, but also in motion—starting with the Gmünd, which had only slightly north of 25 horsepower, to the other end of the line, the ultrasuccessful RS Spyder from the ALMS series. In between them was the 917 Can-Am car that had 1,100 horsepower in its day. It took careful balance to get all these cars on the move at the same time—an achievement I marveled at. Those wonderful designs all dancing in front of my lens, turning on command to separate themselves from the pack! Four cylinders to ten cylinders to even twelve cylinders: they all powered distinctive shapes driven by design and a philosophy of one vision from a single remarkable family.

This history that is exhibited here shows the evolution of taking on challenges—challenges of motorsport and challenges of regulations, not just in racing, but in selling cars globally. You can see a history of design here that reflects the times and the size of engines and the bodywork that it took not just to control it but to allow it to slice through the air. This is a beautiful collection that spans almost sixty-five years. These designs have influenced our lives and enabled us to go and race in places we might not have imagined. These designs have even influenced a museum to recognize them as art and motivated people like me to do what I do.

Commercial film director and racing driver Jeff Zwart has filmed most of Porsche's commercials over the past twenty years. He has published photographic books on Porsche racing cars and is a seven-time champion at the Pikes Peak International Hill Climb, setting records while driving the latest from Porsche.

Porsche Type 356C Cabriolet, 1965, Collection of the Joplin Family, Courtesy of the Rock and Roll Hall of Fame and Museum, Cleveland, Ohio. Photograph © 2013 Peter Harholdt

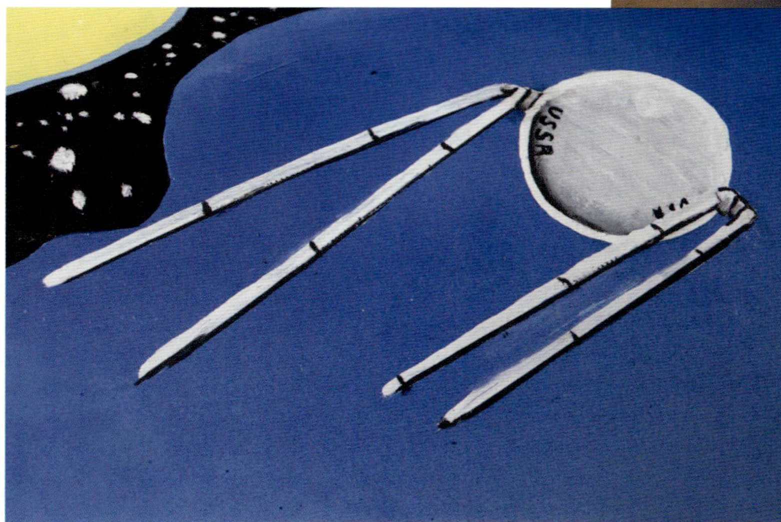

Porsche Type 356C Cabriolet (details), 1965, Collection of the Joplin Family, Courtesy of the Rock and Roll Hall of Fame and Museum, Cleveland, Ohio, Photographs © 2013 Peter Harholdt

Technical Notes

Engine Type: 1,582 cc, air cooled, horizontally opposed, OHV 4-cylinder

Horsepower: 88 hp at 5,200 rpm

Transmission: 4-speed nonsynchromesh manual

Top Speed: 110 mph

Wheelbase: 82.7 in.

Suspension: Independent front transverse torsion bars with parallel trailing arms; independent rear swing axle with a single trailing arm on each side

Weight: 2,150 pounds

PORSCHE TYPE 356C CABRIOLET

"Take another little piece of my heart now, baby! Break it!"

This is the most famous Porsche 356C Cabriolet of all time. It belonged to the first lady of rock and roll, the inimitable Janis Joplin. Born in Port Arthur, Texas, in 1943, Janis Lyn Joplin desperately wanted to be a rock and roll star. In 1966 she headed for San Francisco, where she joined the up-and-coming band Big Brother and the Holding Company.

Her timing was perfect. The city's Haight-Ashbury district had become the capital of the hippie movement, and Janis and the Company performed to screaming, dancing crowds at venues such as the Fillmore West, the Avalon Ballroom, and Winterland. She was rock and roll's first diva. There was really no one like her.

"Oh Lord, won't you buy me a Mercedes Benz?" she would screech in that gravelly, unmistakable voice. "My friends all drive Porsches, I must make amends."

By 1968, with success assured thanks to tunes like "Me and Bobby McGee," "Down on Me," and "Cry Baby," Joplin surely had the means for a new Mercedes, but instead she plunked down $3,500 at Estes-Zipper Motors in Beverly Hills for a preowned "oyster white" '65 Porsche 356C Cabriolet. A civilized convertible with roll-up windows, the 356C was the precursor to the contemporary 911. It was a classy car for a woman who transcended class distinctions.

Joplin reportedly loved the Porsche, but its plain exterior had to go. She asked her friend Dave Richards—Company roadie—to paint the car. She didn't give him specific directions. Richards probably wouldn't have listened. The result was a flowing, Day-Glo, psychedelic masterpiece of rainbows, jellyfish and birds, butterflies and flowers, alligator skin, colorful flags, and faces. There's even a portrait of Joplin and

her band on the front fender. Richards modestly described his work as "the history of the universe."

Joplin drove the painted car to rock and roll venues, and as it was so recognizable, her fans continually left notes and tributes tucked in doorjambs and under the windshield wipers. One night the Porsche was stolen, and the thief tried to disguise it with gray primer. The car was soon recovered and its fabulous finish restored.

Beaded and bell-bottomed, backed by her own groups, Kozmic Blues Band and later the Full Tilt Boogie Band, Joplin performed at the Monterey Pop Festival and Woodstock and starred with the Grateful Dead. She sang her heart out, partied hard, drank to excess, and was swept away. She died of a heroin overdose in a Hollywood motel room on October 4, 1970.

Her parents gave the Porsche to her manager, Albert Grossman, who "drove it into the ground," according to Janis's brother Michael, before the family asked for it back. Michael and sister Laura became the car's caretakers in 1973. They had the Porsche refurbished and drove it for many years, but without its psychedelic paint. "We opted for the car," Michael said. "Janis hadn't become an icon, yet."

Meticulously repainted for *Love, Janis*, a play written by Laura Joplin, the psychedelic Porsche is no longer in running condition. Its odometer reads 138,512 miles. It still belongs to the Joplin family. A historic artifact now, it's been on display since 1995 at the Rock and Roll Hall of Fame.

Driver Zora Arkus Duntov speaks with Ferry Porsche (in checkered cap) from a Porsche Type 550 Spyder at the 24 Hours of Le Mans race in 1955. (Porsche Museum Archives)

ALL COLOR PHOTOS: Porsche Type 550 Prototype, 1953. Courtesy of The Revs Institute for Automotive Research. Photographs © 2013 Peter Harholdt

Technical Notes

Engine Type: 1,498 cc, air cooled, horizontally opposed, OHV 4-cylinder

Horsepower: 98 hp at 5,500 rpm

Transmission: 4-speed synchromesh manual

Top Speed: estimated 110 mph

Wheelbase: 82.7 in.

Suspension: Independent front transverse torsion bars with parallel trailing arms; independent rear swing axle with a single trailing arm on each side

Weight: 1,450 pounds

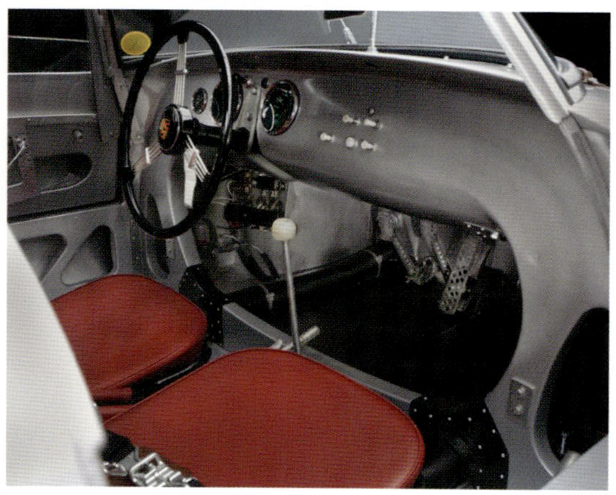

Ferry Porsche (in checkered cap), Hans Klauser (center), and Wolfgang Porsche, the youngest son of Ferry Porsche, during the 24 Hours of Le Mans race in 1956. (Porsche Museum Archives)

Porsche Type 550 Prototype Coupe (chassis 550-001) at Le Mans in 1953; from right to left: Rolf Wütherich, Richard von Frankenberg, Paul Frère (partially obscured), Willi Enz, Huschke von Hanstein, Bruno Trostmann (behind von Hanstein), Hans Herrmann, Hubert Mimler, Ferry Porsche, Hugo Heiner, Wilhelm Hild (in background), Helm Glöckler, Hans Klauser, Ernst Fuhrmann, front (kneeling) Herr Stubbe (Dunlop racing manager). Hans Herrmann and Helm Glöckler achieved sixteenth place overall and won their class. Richard von Frankenberg and Paul Frere drove 550-002. This photo was taken before the race. Hans Herrmann, who was driving his first race for Porsche, did not yet have his racing suit on. Wütherich would later survive the wreck in which James Dean was killed. (Porsche Museum Archives)

PORSCHE TYPE 550 PROTOTYPE

Porsche's success as a sports and racing car manufacturer began with this car, the first of two 550 prototypes. Lacking large advertising budgets, Ferry Porsche knew racing was the crucible in which to prove the marque's superiority and capture world attention.

Work on Project 550, a low-drag racing coupe, began in 1952. Its modified push-rod production-car engine had larger twin-choke Solex carburetors and other modifications. An external oil cooler, wet-sump lubrication system, new transmission with Porsche synchromesh, and limited-slip differential were added. Weidenhausen of Frankfurt built the alloy bodies and removable hardtops for the two cars being readied for the 1953 season.

This car, serial number 550-01, competed as a roadster in its initial race at the Nürburgring, driven by Helmut "Helm" Glöckler. He battled rainy weather, overcame carburetion problems, and won, adding the Porsche 550 to that short list of cars that were victorious in their first race.

At Le Mans that year, driver Hans Herrmann joined Glöckler in 550-01, and a pair of driver-journalists drove 550-02. Both coupes crossed the finish line at the same interval at which they started. Le Mans scorers abhor a tie, so the class win was awarded to 550-02. The racing journalists had won the 1,500 cc class and set a new record.

For years Porsche enthusiasts believed the two 550 prototypes had vanished, but 550-01 survived in a warehouse in Mexico. Its original aluminum shell was worn from competition, so Salvador Lopes-Chavez, who'd raced it in 1954, built a fiberglass body. Later Joe Cavaglieri Restoration of Sherman Oaks, California, crafted a new aluminum body. After a five-year restoration, 550-01 was the Concours de Sport Best of Show Winner at Amelia Island in 2005.

154

UPM
x1000
RPM

DENISE McCLUGGAGE **IN THE DRIVER'S SEAT**

"It'll all come back to you. Regular H pattern shift. Enjoy." Bob Snodgrass is handing over to me the Brumos now-vintage Porsche 550 RS60. It's 1989 at Laguna Seca Raceway, and this is practice time for an exhibition run in celebration of Porsche's fiftieth anniversary. The car is as shining black as a licorice whip and twice as smart. For all its vintage status, it is a year newer than any of the 550s I had raced in prime time, mostly for Briggs Cunningham.

I go out on the Laguna course. H-shaped shift, huh—so why am I grinding these gears like a dental college dropout? Then midway into the first inept lap I remember: Ah, a *script* H. Script! And I extend the gear lever far up to the right, and it did indeed all come back: the extremely wide throws on the Porsche gearboxes of the era. Not a proud feature. And I recalled how much more snick-snick the lovely gears in the Maserati brothers' lovely OSCAs were. So why not stick with OSCA? Simple. Porsche was whipping OSCA's snick-snick like a fairy-tale stepmother. Porsche was tellingly faster, And *fast*, as Phil Hill once spelled out for a journalist, is the whole intention of the thing.

It came to me during the rest of my now-delightful ride in the licorice 550 RS60, Porsche's entire plan of development that

Porsche Type 718 RS60, 1960, Bowman Motors, Photograph © 2013 Michael Furman

By 1957, nine years after the first Porsche 356 was handcrafted in Austria, Porsche was celebrating over 700 international racing successes with championship wins in France, Belgium, Holland, Switzerland, Germany, Austria, Denmark, Spain, and the U.S.A. in this poster designed by Erich Strenger. (Porsche Museum Archives)

continues today with constant tweakings even in the midst of model runs. The scheme as I read it: change first what matters most, then constantly fill in. Fussbudgeting, improving. Repeat.

The engine size of the 550s I raced for Briggs was at the largest, I think, 1,500 ccs. Barely a respectable touring motorcycle. Briggs almost pleaded with the factory: "Build at least a two-liter. We could be winning overall." Time showed he was not privy to the schemes that dwelt in the heart of Stuttgart. I never even drove an RSK, never mind the coming behemoths that began playing Pac-Man with Le Mans competitors and set surely unbreakable records over the years.

Frankly, winning wasn't the most important thing with me. Though anyone who keeps telling you that never *won* anything. Which I discovered after a few victories. But nonetheless what I loved was the process, like skiing. I loved the physicality of turns well executed. And I loved a neat, surprise pass laid on a bigger car. It was the aesthetics of the kinesthetic that delighted me in racing. And the Porsche 550 was a terrific dance partner. A happy synchronicity. I'm glad I raced in its era.

I had always said that I loved my Italian cars but I admired the Germans. Though soon I noticed I couldn't get out of the 550 without trailing a hand along the swell of a fender. *Ich liebe dich auch*. The love affair played out in many venues. Cumberland, Marlboro, Virginia, Elkhart Lake, Bridgehampton, Watkins Glen, Nassau, Venezuela. Maybe others.

Watkins Glen led to a pleasant ego tickler: a plaque with my name on the downtown sidewalk. Nassau is remembered by some older few as the best ladies race ever. Certainly one of the few ever watched by the men. And Cumberland I always thought was my best run anywhere but with a typical disappointment as a cap.

First Nassau. Ruth Levy was the hot female shoe on the West Coast, and I raced on the East. The competitive nature—surprise—of the male contingent stirred up "our girl can beat your girl" childish taunts. To force some growing up, I invited Ruth to be my guest at Nassau Speed Week and to meet in the ladies race to discover some truths. We were both to be in Porsche 550s and were both familiar with the breed. But in the event, as the Brits say, Ruth drew the appreciative eye of an unwed Stirling Moss, who persuaded John Wyer, owner of the Aston-Martin Stirling was to drive in the "real" races, to lend Ruth that car. So the match race turned into 1,500 ccs versus 3,700 ccs.

In the first heat, Ruth was bouncing off barrels and having a fine handful of a time. I managed to win by something more than a nose. In the second heat, I had again managed a lead going into the final turn, a large U. Always heavy of foot, Ruth was growing in my side mirror with incredible insistence. Wha'? All she had to do was follow me into the turn and use her power to blast by and erase my few-inches advantage earned in Heat One. But no. On Dancer! On Aston! Oh boy, I'm not turning in front of *that*. And Ruth didn't bother with the turn, either. Off the road's end and rolling in the gorse. I tiptoed home the winner and blasted back to the turn of disaster. My speeding roommate was off to the hospital and, they reported, was OK.

At Cumberland I drove what may have been my best race ever. Won the ladies race and was in the big boys race driving with aplomb, a favorite companion. I passed

Porsche Type 718 RS60, 1960,
Bowman Motors, Photographs
© 2013 Michael Furman

An English-language version of this classic Erich Strenger poster celebrates quadruple Porsche successes at the 1959 Targa Florio. (Porsche Museum Archives)

everything my size and some that were bigger, finished high overall, and felt tall enough to play center. But then a stick-and-ball journalist, whose type usually covered races then, wrote that I won the ladies event but "against the men could only manage fifth." Or maybe it was sixth. Hey, Boobie, I was *first* in class. And the unmanageable ahead of me were D-Type Jags and Listers and other honking things. The Porsche was terrific, you ignorant misogynist.

(So that's off my chest.)

The Porsche 550s inevitably went under tarps in dusty corners awaiting a renaissance in the vintage craze. I did do vintage a few times, but owners too soon discovered the fun that awaited the do-it-yourselfer. Anyway, racing was not my career, writing was. And as the wise man advised: "Shoemaker, stick to your last." I did. But I'd have given it all up for a Porsche.

Denise McCluggage raced when vintage cars were new and wrote about and photographed the cars and their drivers. She has written for AutoWeek *forever and has a Web site www.denisemccluggage.com. She is converting her books and others with car themes into e-books for Fulcorte Press.*

Technical Notes

Engine Type: 1,587 cc, air cooled, horizontally opposed, DOHC 4-cylinder

Horsepower: 178 hp at 7,800 rpm

Transmission: 5-speed manual transaxle

Top Speed: 142.9 mph

Wheelbase: 86.6 in.

Suspension: Independent trailing-arm front suspension with torsion bars; independent double-wishbone rear suspension with coil springs

Weight: 1,280 pounds

PORSCHE TYPE 718 RS60

Porsche's first purpose-built racing cars were little more than aluminum eggshells on wheels. As aerodynamics improved and engine output increased, each successive competition model became faster and more agile.

The Type 718, a lightweight two-seater also known as the RSK Spyder, was introduced in 1957 as a successor to the race-proven 550A/1500RS. The mid-engine RSK shared the 1500RS's space-frame structure, along with its wheelbase and track dimensions, but its lightweight alloy body was narrower and, some say, prettier. Aerodynamic drag was reduced. Stability at high speeds improved further in 1958, when a few cars were fitted with small stabilizing fins on the rear fenders.

As a result of FIA (Fédération Internationale de l'Automobile) regulation changes in 1960, the model 718 RS60 had a four-inch-longer wheelbase and a more extended nose section than those of its predecessors. The RS60's frontal aspect may have influenced Carlo Abarth's later design for the Carrera GTL. The new RS60 conformed to FIA rules that required legal road equipment, including deep bucket seats, a tall windshield, a folding top, and nominal space for luggage. Its quick-release fuel tank cap was offset to the right; early 550s had the cap in the center.

An ancestor of the contemporary Porsche Boxster, the RS60 was the ultimate development of Porsche's four-cam Spyder. The RS60 joined the ranks of Porsche's "giant killers" when a racing Spyder with a 1,679 cc engine upgrade, driven by Joakim Bonnier of Sweden, Hans Herrmann of Germany, and Graham Hill of England, won the 1960 Targa Florio in Sicily, achieving more than a six-minute lead over a more powerful three-liter Ferrari.

This model is one of only twelve to fourteen customer RS60s built. Original owner Bill Wuesthoff placed in the top three in the under-1,500 cc class in seventeen races and won in class an impressive twelve times in four years.

Ferry Porsche (in hat) at the 1961 24 Hours of Le Mans race in France. Against competition like the 3.6-liter Aston-Martin and the 3-liter V-12 Ferrari, Porsche drivers Masten Gregory and Al Holbert in a 2-liter Porche Type 718 Spyder RS61 achieved an impressive fifth place finish overall. Edgar Barth and Hans Herrmann achieved seventh place in a 1.6-liter Type 718 Spyder RS61, and Ben Pon took tenth place in a Porsche Type 356 1600 GS Carrera GTL Abarth Coupe. Out of 55 starters, only 22 cars finished the race. (Porsche Museum Archives)

Porsche Type 804 Formula One, 1962, The Porsche Collection of Ranson W. Webster, Photographs © 2013 Peter Harholdt

ABOVE: American driver Dan Gurney in August 1975 in the air-cooled six-speed Porsche 804 Formula One car in which he won the French Grand Prix in 1962. (Porsche Museum Archives)

PORSCHE TYPE 804 F1

As a tiny company with limited resources, Porsche hadn't planned to race in European Formula One Grand Prix (F1), but a fortuitous rules change made all the difference. F1 racing was expensive, and the technology had little to do with Porsche's road cars. Porsche had been competing in Formula Two because the engine size was similar to that of its road cars, but in 1961, when the F1 engine displacements dropped to 1,500 cc, Ferry Porsche decided to plunge ahead.

Porsche decided to build its own engine and steel space-frame chassis to stay competitive. The all-new flat eight had a magnesium crankcase and alloy twin-cam heads. Porsche competed for the first time at Zandvoort, in the 1962 Dutch Grand Prix. The drivers were Sweden's Joakim Bonnier and America's Dan Gurney. Neither did very well. Ferry Porsche was keenly disappointed, almost to the point of ending the F1 effort. But after extensive chassis development and engine work, Gurney's 804 won Porsche's first Grand Prix race, in France. Soon afterward, at Solitude, Gurney won the race after setting a new lap record. Bonnier was second. That 1–2 finish was the high point of Porsche's F1 experience.

After Gurney finished third at the Nürburgring, the F1 car was made lighter and its aerodynamics improved, but mechanical maladies plagued the Porsche 804s. Porsche did not enter the last race that year, held in South Africa.

Porsche began making improvements to its F1 engine and chassis for 1963, but the anticipated high cost of the F1 program caused Ferry Porsche to release Gurney and Bonnier and elect not to continue in F1. The flat eight was enlarged and used in various forms until 1968 in the Type 907 chassis.

This Type 804 is the winning car that Dan Gurney drove at Rouen and Solitude. Two other Type 804s survive, one in the Porsche Museum and the other in the U.K.'s Donington Collection.

Technical Notes

Engine Type: 1,494 cc, air cooled, horizontally opposed, carbureted DOHC 8-cylinder

Horsepower: 180 hp at 9,200 rpm

Transmission: 6-speed manual transaxle

Top Speed: 168 mph

Wheelbase: 90.5 in.

Suspension: Independent front double wishbones, longitudinal torsion bars, inboard-mounted shocks; double rear wishbones, torsion tube, inboard-mounted shocks

Weight: 996.5 pounds

Porsche Type 804 Formula One, 1962, The Porsche Collection of Ranson W. Webster, Photographs © 2013 Peter Harholdt

LEFT: Dan Gurney at the wheel of a Type 804. (Porsche Museum Archives)

CAMERON HEALY

THE GLORY OF GOODWOOD

Editor's note: Many Porsche enthusiasts long to own a factory race car. A few lucky individuals have achieved that goal. We can't travel back in time, but a second fantasy would be to compete today in a historic Porsche team car. Cameron Healy, a lender to this exhibition, owns several vintage Porsche competition cars. In this essay he tells what it's like to compete at the annual Goodwood Revival—and do so quite convincingly—in a Porsche factory team car.

Bicycling through the Laguna Seca Paddock during the Monterey Motorsports Reunion in August 2012, I was waved down by Andy Prill of the respected UK Porsche shop Maxted-Page and Prill, who'd been asked to find a replacement Porsche 904 to participate in the Goodwood Revival, just one month away. I applied and quickly received confirmation from Goodwood. We made arrangements to ship the Porsche to England, and before long, it was time to race.

a warm-up lap, stopped at our grid positions (with the '60s-attired Popsicle girls marking our respective grid spots), then practiced doing a mock start. After a five-second warning and engines revving, the Union Jack flag was dropped, and throttles went hard down, creating a smoky cacophony. I got a good start and made up a few places before drifting wide into Madgwick Corner.

I let the most aggressive car by to create some breathing room, then tore through the two right-turn apexes and shifted into fifth gear at full throttle with a slight lift for Fordwater. Then it was fully back on the power until hard braking and a downshift to fourth for the very quick St. Mary's right turn and an off-camber left. Next it was back on the power for a drag race to Lavant Corner, taking its two right-hand apexes in third gear. You must carry speed onto the long Lavant Straight. Back up to fifth, we were hitting 130 to 135 mph before preparing to brake and downshift for a late apex at Woodcote Corner, followed by a tight right-and-left kink in front of start/finish. Then you do it again for another sixteen laps.

After my initial lap, I felt more comfortable and began passing cars. As I attempted to overtake, they would move over in front of me in a blatant and classic effort to block my access to an improved position. I tried feinting one way, then abruptly moving the other way. This tactic created just enough of a delay to hold me back until I could outsmart each opponent. But with every lap, I found confidence, and soon my lap times were coming down. I saw 1:37s and

ABOVE: Targa Florio 1969. Ferry Porsche in discussion with drivers, from left to right: Umberto Maglioli, Dick Attwood, Brian Redman, Ferry Porsche, Hans Herrmann, Udo Schütz, Rolf Stommelen, Vic Elford, Rudi Lins, Gérard Larrousse. At the wheel of Porsche Type 908/02 with starting number 266 is Gerhard Mitter. The Targa Florio 1969 was one of the biggest victories for Porsche: First Place: Gerhard Mitter / Udo Schütz, Second Place: Vic Elford / Umberto Maglioli, Third Place: Hans Herrmann / Rolf Stommelen, Fourth Place: Karl von Wendt / Willi Kauhsen, all in Porsche Type 908 Spyders. (Porsche Museum Archives)

LEFT: Targa Florio 1968. Vic Elford's Targa Florio victory driving a Porsche 907 was so decisive in 1968 that Porsche for the first time put a driver front and center on its celebratory poster. Elford wrote, "Porsche racing engineer Helmuth Bott told me that for once, Porsche decided it was the driver who won the race—not the car." (Porsche Museum Archives)

Porsche Type 904/6 Prototype, 1965, Private collection of Cameron Healy and Susan Snow, Photograph © 2013 Peter Harholdt

PORSCHE TYPE 904/6 PROTOTYPE

Technical Notes

Engine Type: 2,000 cc, air cooled, horizontally opposed, DOHC 6-cylinder
Horsepower: 245 hp at 8,000 rpm
Transmission: 4-speed synchromesh manual
Top Speed: 160 mph
Wheelbase: 90.6 in.
Suspension: Independent front double wishbones with coil springs; independent rear multilink with coil springs
Weight: 1,443 pounds

Porsche's Type 904 succeeded the RSK Type 718 as the last sports-racing iteration of the 356 Series. Developed after Porsche left Formula One in 1962, the 904 was also the last full-competition Porsche that could be readily street-driven.

A ladder chassis embraced the mid-mounted alloy four-cylinder, four-cam Carrera engine. The 904 borrowed its suspension design from Porsche's F1 car. Its steering, adapted from the just-launched 901/911, was a new ZF rack-and-pinion design. To save weight the body was fiberglass (made by spraying chopped fiberglass onto a mold), and it was bonded to the chassis for added rigidity. The bucket seats were fixed, but the pedals were adjustable. Porsche made several seat sizes available to suit different driver heights.

Porsche 904s nearly always finished right behind the leading big-bore race cars, and their success paced Porsche's return to the forefront of international competition. In Sicily's 1964 Targa Florio, a pair of 904s finished ahead of the Ferraris and Shelby Cobras. For 1965–66, a series of engine improvements including bigger valves and reprofiled camshafts made the 904 even more competitive.

In 1965 twenty cars were fitted with six-cylinder engines (904/6), and even a few eight-cylinder Type 771, 225 hp, two-liter cars (904/8) were built. This car is one of six two-liter, six-cylinder prototypes constructed in 1965 as Porsche factory racers. Only four survive. This car, 906-011, was driven by Gerhardt Mitter in the 1965 Le Mans trials. That same year, it finished ninth in the 1,000 km of Nürburgring and won third place at the Grand Prix of Solitude in Stuttgart.

In 1968 Porsche AG sold this car to Vasek Polak in California, without an engine. With a four-cylinder, four-cam engine installed, it competed at tracks from Riverside, California, to Watkins Glen, New York. During a restoration in 1985, a correct two-liter, six-cylinder engine was fitted. This car has been campaigned in vintage racing events.

Porsche Type 904/6 Prototype, 1965, Private collection of Cameron Healy and Susan Snow, Photograph © 2013 Peter Harholdt

then 1:36s on my pit board, and then I was into the 1:35s. My goal was to get closer to the other 904/6 (a sister car to mine) driven by Irvine Laidlaw, who's quite experienced at Goodwood. My car (chassis number 906-011) was used by the Porsche factory team for the 1965 Le Mans trials in April of that year, while Irvine's car (906-012) was the one used in the actual 24 Hours of Le Mans that June. He had qualified in the high 1:34s.

At a little more than the halfway point, I tried to engage the clutch and discovered I had no pedal. The missing pedal was lying on the floor, face down. The peg that holds the pedal up had given way; I'd have to adapt, or my race was toast. With the tip of my toe, I would move the pedal into position to up- or downshift, but I would have to let it fall back on the floorboard for lack of leg strength to hold it up and the need to use my left foot for bracing as I went around corners. This slowed my progress, and I was lapped by a car I had passed earlier. Distracted, I swung wide with two wheels on the grass entering Lavant Straight. Soon I figured out a rhythm and was back into the 1:36s and 1:35s.

Far too soon, the checkered flag came down, and the race was over. I finished nineteenth out of twenty-seven cars. As the first five placing cars had had professional drivers, I felt pretty good. When I pulled into the holding area, I began tuning in to the sights and sounds of Goodwood, my senses keen because of the heightened concentration that racing brings. With Spitfires and Hurricanes flying overhead, old-time music playing, the sound of the enthusiastic crowds, and the throaty roar of old race car engines, I felt as though I had been pulled back into a time warp of many decades past.

There are plenty more stories to tell about watching all the great racing over the three days, as well as Lord March's mind-blowing Dr. Zhivago–themed party in a transformed airplane hangar on Saturday night. There were camels and a North African World War II scene at the drivers club; the wonderful Dan Gurney tribute on Sunday; lunching on Sunday with Danny Sullivan, Jeff Zwart, and his lovely spouse, Terri; Suzy and my wind-down dinner on Sunday evening at the Goodwood Hotel, sitting at an adjacent table to Stirling and Susie Moss on the eve of Stirling's 83rd birthday ... and on and on I could go. These stories will need to be saved for another time.

For four days we were transformed and eased into another time of style and grace, when sportsmanship had a different meaning. And by going into the past, we were brought to an enhanced place in the present, allowing us to emerge back into our normal lives with fresh perspectives and an elevated appreciation for all that is good.

Thank you, Goodwood and Lord March!

Cameron Healy has been a passionate Porsche enthusiast since age fifteen and currently owns and races a small but quality stable of historically significant classic Porsche racing cars in the U.S. and Europe.

ABOVE: Author Cameron Healy driving the 904/6 at the 2010 Le Mans Classic, Photograph © 2013 Cameron Healy

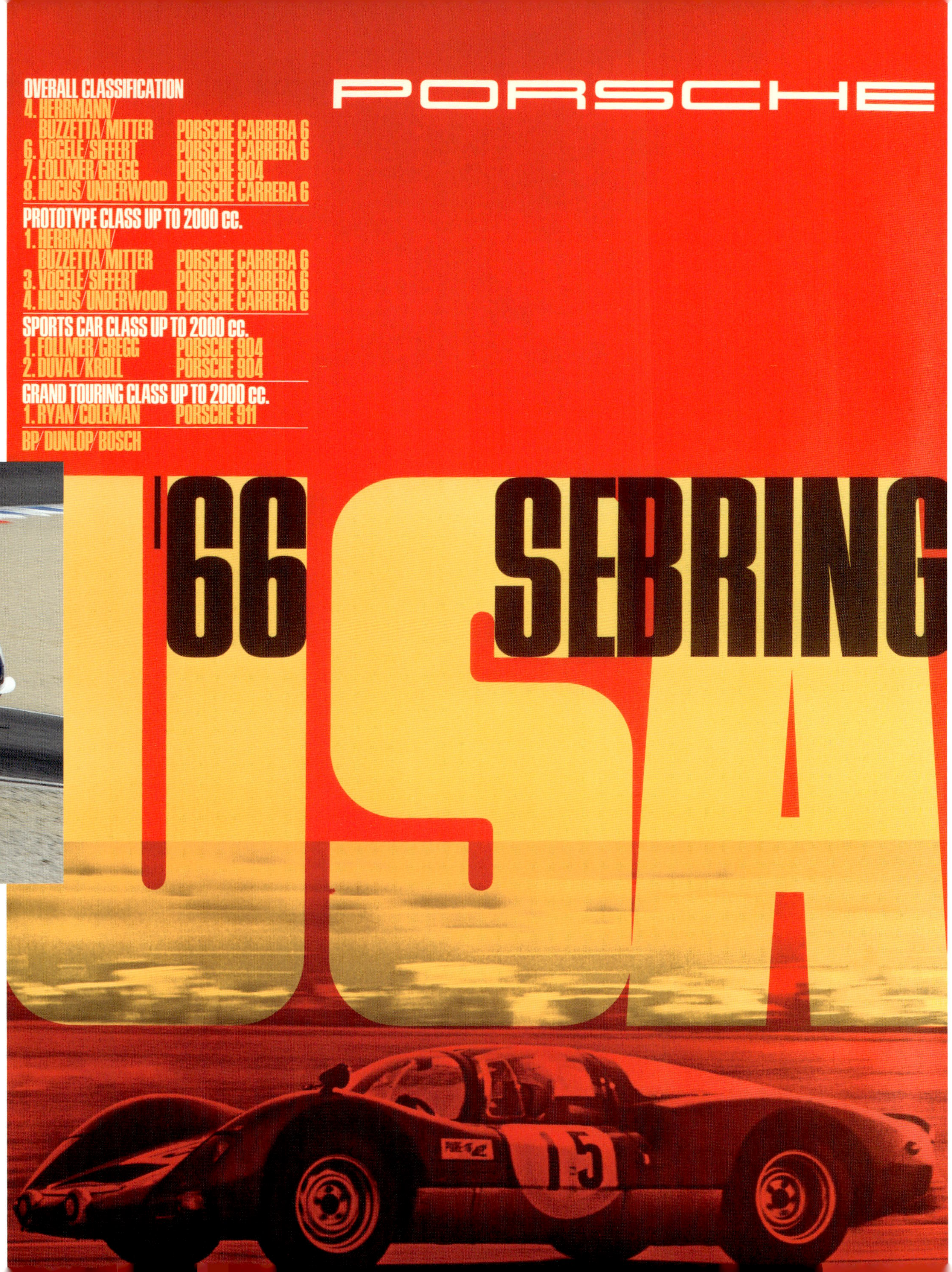

ABOVE: Cameron Healy drives the Porsche Type 908K Prototype in 2009 at the Monterey Historics at Laguna Seca Raceway. Photograph © 2013 Hal Thoms

RIGHT: Vintage Porsche poster celebrating motorsports successes at the 12 Hours of Sebring endurance race in 1966. (Porsche Museum Archives)

The Porsche Type 908K Prototype featured in the NCMA exhibition (car number 6 in foreground of this photo) racing at the 1,000 Kilometers of Spa-Francorchamps race in Belgium in May 1968. (Porsche Museum Archives)

Porsche Type 908K Prototype, 1968, Private collection of Cameron Healy and Susan Snow. Photographs © 2013 Peter Harholdt

6

THIS PAGE: Porsche Type 908K Prototype, 1968, Private collection of Cameron Healy and Susan Snow, Photographs © 2013 Peter Harholdt

RIGHT: Ferry Porsche with nephew Ferdinand Piëch (left) and sons Hans-Peter and F. A. "Butzi" Porsche. (Porsche Museum Archives)

Technical Notes

Engine Type: 2,990 cc, air cooled, horizontally opposed, fuel injected DOHC, 16-valve, 8-cylinder

Horsepower: 350 hp at 8,500 rpm

Transmission: 6-speed manual transaxle

Top Speed: 200 mph

Wheelbase: 90.6 in.

Suspension: Independent front double wishbones, coil springs; upper and lower rear links, radius arms, coil springs

Weight: 1,322.8 pounds

Porsche was always ready with a new race car when formula requirements changed. The company had set its sights on an overall victory at Le Mans, but the engine size rules were not in its favor. That would change in 1968, when the FIA (Fédération Internationale de l'Automobile) regulations for Group 6 Prototype sports cars limited displacement to 3,000 cc.

The 908 became the first Porsche sports car with an engine that was the maximum size allowed for the Prototype class. Developed by Hans Mezger and his team,

PORSCHE TYPE 908K PROTOTYPE

the 908 engine was based on the six-cylinder 916, with two additional cylinders. The first 908Ks, like this car, were configured as closed coupes with minimum aerodynamic drag, suitable for high-speed tracks. They had wider bodies, fifteen-inch wheels, larger fuel tanks, and short (*Kurzheck*) tails. A 908K driven by Jo Siffert and Vic Elford won the Nürburgring 1,000 km in 1968.

To increase speed, new 908Ls (*Langheck*) were fitted with longer tail sections, but these cars were unstable at high speeds, so Porsche engineers designed a movable suspension-controlled automatic rear flap (called an elevon) to reduce instability. Modified 908L Porsche cars led in early stages of the 1968 Le Mans 24-Hour race, but electrical troubles and eventual disqualification marred Porsche's entry. A Ford GT40 won; a 907 was second, and the 908K was third.

This car, 908-010, was a Porsche factory race car, and it's thought to be one of five short-tail prototypes extant. It was one of two examples raced at the 1,000 km of Spa in May 1968, co-driven by Vic Elford and Jochen Neerpasch. Elford drove thirty-two laps, and then turned the car over to Neerpasch, who managed two laps before skidding off the course into a telegraph pole. This car did not race again for Porsche and was sold to a Swiss privateer. In the late 1990s it was sold by Bill Ferren to Dale Miller and underwent a complete restoration, completed in time for the 2004 Rennsport Reunion at Daytona. The present owner races the car in vintage events.

PORSCHE
AUDI
GOODYEAR
PORSCHE
AUDI
GOODYEAR

Mark Donohue in number 6 Porsche 917/10K leading Penske teammate George Follmer at Donnybrooke Can-Am 1972. Photograph © 2013 Pete Lyons

PETE LYONS RETHINKING SPEED

When a sculptor—let us think of Rodin—begins to develop a new work—*The Thinker,* if you will—many conflicting factors and forces must be evaluated, balanced, and resolved. Dimensions, proportions, internal dynamics, technologies, materials, weights, means of construction, surface treatment, display space, audience ... any number of mundane considerations arise.

But always, the primary thrust is spiritual. The artist strives to express a fresh understanding of an emotional truth.

So it is when a constructor of racing cars—Porsche, say—creates a breakthrough combat machine. The thinking behind myriad engineering choices is aimed toward a single truth: the fastest race car.

I used to ponder this, without really knowing I was, early in my career as a motorsports reporter, when I lived near Philadelphia and often drove past *The Thinker* poised outside that city's Rodin Museum. This is a powerful piece, a revolutionary artist's depiction of a muscular man whose torso is torqued to express invisible mental intensity. It makes you think.

Mark Donohue contemplating the complex turbocharged engine in his Porsche 917/10K in the pit lane at the 1972 Riverside Can-Am. Photograph © 2013 Pete Lyons

Apparently this unrealized insight remained in my sub-brain, ready for the day I stood, dumbstruck, before Porsche's Can-Am race car of 1972.

It was the most powerful racing sports car the world had ever seen. Almost a thousand horsepower! A magnificently muscular machine, it seemed twisted internally by the terrible torque that twin turbochargers unleashed from its twelve cylinders.

Turbocharging is the art of using an internal combustion engine's own fiery exhaust gases to compress fresh induction air, the "charge," thus cramming more potential power into itself. It's as if the engine lifts its own performance, bootstraps-style. Turbocharging is supercharging, but done with turbine wheels operating in a gaseous medium rather than being driven directly through gears or belts.

Bolting on a turbo or two is an efficient way to boost power, and manufacturers of trucks (and aircraft) were doing it early in the twentieth century. For automobiles, though, turbocharging is trickier, especially in race cars used on road circuits. The problem is called throttle lag. It means that when the driver moves the gas pedal, there's a delay before the engine responds.

Many auto racing experimenters before Porsche had tried turbos, and they'd done well enough in Indycar speedway racing, where the engine normally is kept at high power. But in road racing, speeds vary greatly around the track. Nobody had worked out how to give the driver, dancing on the feathery edges of control through many different curves, the necessary delicacy of response to the accelerator pedal.

Think of burly Monsieur Rodin trying to sculpt with tools at the ends of long, rubbery wands.

Can-Am was short for Canadian-American Challenge Cup, a North America-wide, professional road racing series inaugurated in 1966. The Can-Am was exceptional for imposing very few restrictions on car technology. Contrary to most motorsports formats, even way back in the 1960s, in the Can-Am engine size and design, chassis structure and dimensions, aerodynamic conception, car weight, fuel tank capacity, tire width ... all were basically free. As long as your contender had bodywork covering the wheels and a sketchy passenger seat, so it could be described as a "sports car," scarcely anything else was out of bounds. The Can-Am was an inventive experimenter's paradise.

America being an important market, Porsche wanted to race in the high-profile Can-Am Series. The German automaker made exploratory efforts in 1969 and again in 1971, but even its biggest engine, a 5.0-liter (302 cubic inch) twelve-cylinder designed for Le Mans endurance racing, was two-thirds the size of the big-block Chevrolets in the dominant cars.

Engine chief Hans Mezger.
(Porsche Museum Archives)

Porsche's cars were reliable, but they were slow. Not the impression the company wanted to make.

Engine chief Hans Mezger saw two ways forward. First, his engine shop simply added four cylinders to the twelve, making an experimental sixteen. The increased displacement did yield plenty of power, but the sprawling

Technical Notes

Engine Type: 7,200 cc, air cooled, horizontally opposed, DOHC 16-cylinder

Horsepower: 880 hp at 8,300 rpm

Transmission: 4-speed synchromesh manual

Top Speed: 225 mph

Wheelbase: 100.6 in.

Suspension: Front and rear unequal-length upper and lower A-arms, coil springs

Weight: estimated 1,850 pounds

Porsche Type 917 16-cylinder Spyder Prototype, 1969, at the Porsche Museum, Courtesy of the Porsche Museum, Photographs © 2013 Art Howard

The Type 917 race cars, designed for European endurance contests, gave Porsche its first overall Le Mans wins in 1970 and 1971. But the popular Can-Am races in North America were usually 200-mile high-speed sprints. Big-block Chevrolet-powered McLarens dominated this exciting series.

PORSCHE TYPE 917 16-CYLINDER SPYDER PROTOTYPE

Dr. Ferdinand Piëch, head of Porsche's racing division, felt that a larger-displacement, high-revving engine could beat American big-block V-8s in Can-Am competition. Hans Mezger, whose design team developed the 917 racer, reengineered a Type 912 engine, adding four cylinders. The 180-degree flat sixteen shared the twelve's cylinder head design, but the inlet ports and ram tubes canted outward to ensure room for the belt-driven Bosch fuel injection pumps. Porsche engineers planned for sufficient components to build ten engines. The normally aspirated engines each weighed 706 pounds, thanks to the use of aluminum, carbon fiber, and titanium.

The flat sixteen never ran in competition, but it was tested extensively. One tester was race driver and engineer Mark Donohue, who described the experience in his book *The Unfair Advantage*:

At Weissach's test track in 1971, "A truck showed up with another 917. It looked like the Spyder we had seen before at Can-Am races...except that it was longer. It turned out to have a sixteen-cylinder motor that they wanted me to try...I drove the thing for them, and it was a real monster. The motor was so long you could hear one end start up before the other...Although it wasn't set up properly for any lap records, it was truly impressive in a straight line. They just wanted me to feel the motor, anyhow, and when we were through, they stuck it back in a barn somewhere."

The 917's twin-turbo flat-twelve engine was physically smaller than the sixteen, but it eventually produced even more power. After it was developed fully, it won the Can-Am Series in 1972 and 1973. Rarely seen, the sixteen is an intriguing sidebar in the Porsche racing story.

monster of a motor was hardly efficient or easy to package. Nor was just adding pistons much of an engineering challenge.

More appealing was the second option, to turbocharge the twelve. The fact that nobody in road racing had perfected such an installation meant the prestige accruing to the small, boutique automaker would be all the greater, if its engineers could pull it off.

They almost didn't. Through many months of painstaking fabrication and testing, engineer Valentin Schaeffer and his team evolved an extremely complex system of air ducts, pressure relief valves, and interconnected linkages. The twin-turbocharged Can-Am engine bristled with immense power. But not drivable power. It just would not respond smoothly on the factory's twisty test track.

Toiling alongside the Porsche men was American Mark Donohue, a championship-winning race driver whose degree from Brown University gave him matching credentials as an engineer. On May 27, 1972, Donohue won the Indianapolis 500 in a turbocharged speedway car. Two weeks later, at Mosport Park in Canada on June 11, he would debut the new turbocharged Porsche sports car in the Can-Am series—if it were ready.

He didn't think it would be. Anguished, he advised his Philadelphia-based team boss, Roger Penske, to prepare to withdraw the car. That would be a public relations disaster for all involved, but it would be worse to be there and be slow.

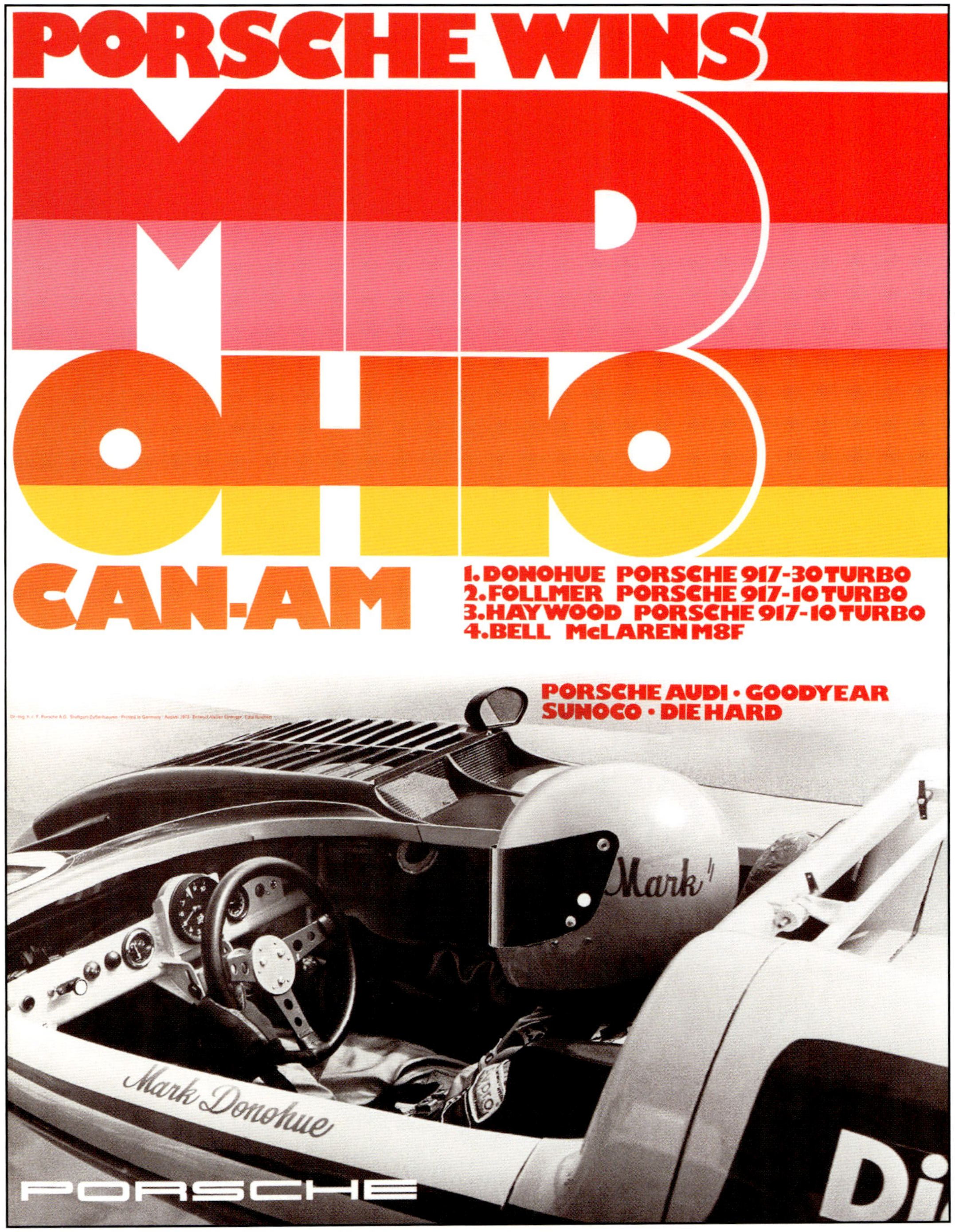

Donohue endured one more flight to Stuttgart, desperate to make one last effort to coax sufficient low-speed throttle response from the engine. Fancifully, I visualize him slumped in his airline seat, *The Thinker* tormented by those damned turbochargers.

It is history that, at absolutely the last minute, Mark Donohue and the Porsche engineers came up with new calibrations for the intricate fuel injection system, and—jubilation!—the engine came right.

All that remained was to prove it on the racetracks. Racing being racing, the road was long, hard, and painful; Donohue spent weeks in agony after a savage crash injured his legs. It was even more agonizing for him to see Penske's stand-in driver, George Follmer, become the 1972 Can-Am champion in the car that Donohue worked so hard to perfect. He said later, "It was like watching someone make love to your wife."

Donohue did recover and in 1973 finally won the Can-Am title with a faster, even better edition of the turbocharged Porsche. This one packed well over a thousand horsepower; if necessary, the driver could increase the turbo boost and dial up as much as 1,500!

Porsche went on to apply its turbocharging expertise to many of its road cars, which, of course, have become renowned for very high performance. Meanwhile, it continued to advance turbocharging in other racing arenas. Successful results included numerous victories in the twenty-four-hour endurance race at Le Mans, plus three drivers' world championships in Formula One. Incredibly, Porsche's tiny F1 engine, a six-cylinder displacing just 1.5 liters, squeezed out power equivalent to the old Can-Am twelves, which had been more than three times the size.

Today, turbocharging is becoming ever more useful to the mainstream industry as it strives to meet goals for "green" vehicles. Porsche is not the only automaker whose work made this difficult technology viable, but it was the historical leader, and that is because of racing. It was the company's drive to perfect turbocharging for its Can-Am campaign that made it happen fast.

Those epic racing machines are museum pieces now, still and silent sculptures, recordings frozen in metal of the revolutionary thinking that made history.

Motorsports writer-photographer Pete Lyons covered Porsche's revolutionary Can-Am championship in 1972, and three of his books are about that technology-changing, no-holds-barred racing series. Learn more about his work at petelyons.com.

Hans Mezger (in suit) and colleagues with the 1983 Porsche TAG-Engine. (Porsche Museum Archives)

Jules
SIMPSON
MOET
Jules
Jules

DEREK BELL **TO PORSCHE, WITH LOVE**

I have been so fortunate to have driven race cars for Porsche, an experience that really made my career. Without the opportunity I was offered back in 1970, and without those wonderful engineers and mechanics, my life would have been very different, and I certainly would not have been invited to write these few words!

In 1968, while I was under contract to Ferrari for Formula One and Formula Two, David York, team manager at John Wyer Automotive, who was then running Gulf GT40s, invited me for a test drive at Thruxton with a view to driving for them at Le Mans that September (the race had been delayed by student riots in June). It was my first taste of sports cars. I assume it went off well, as I was told, "That was very good; you will share with Rodriguez."

Author Derek Bell (left, in red) with co-driver Jacky Ickx (middle, in white) and Ferry Porsche (right) celebrating their first-place victory at Le Mans in July 1981. (Porsche Museum Archives)

I naively asked Enzo Ferrari if I could drive for Gulf Ford—not the cleverest of requests. Il Commendatore's curt reply was: "Honor your contract. (signed) Ferrari." I had no idea about the Ford-versus-Ferrari dramas occurring behind closed doors at that time: I was simply a driver! So that ended what could have been my first Le Mans victory, as Pedro Rodriguez went on to win.

My first experience in a 917 was driving in the *Le Mans* movie with Steve McQueen in 1970. That year David York phoned once more and asked me to test at Goodwood with a view to driving the magnificent Porsche 917 for John Wyer. Ronnie Peterson and Peter Gethin tested with me, and somehow I took the drive for 1971; I'm still not sure why, but Ronnie was on his way to Formula One and (fortunately for me) wanted to concentrate on that.

That fall I had my first experience with Porsche Engineering in Weissach, when we tested aerodynamics for Le Mans in pouring rain at the Hockenheim circuit, with Norbert Singer and the engineers. That resulted in the most terrifying moment I had in all my racing years. Late one afternoon, under dull and wet conditions, entering what are called the Senna curves, flat out at over 150 mph, I suddenly burst through the gloom to see this poor man bicycling across the track. There was no bump, so he must have made it. His shocked expression has been stuck in my memory ever since. The track was obviously his shortcut home

FAR LEFT: During the filming of *Le Mans* in 1970, from left to right: Steve McQueen, Derek Bell, Herbert Linge (back to camera), Gijs van Lennep, and David Piper. (Porsche Museum Archives)

LEFT: Weissach's Porsche Engineering Team colleagues Helmuth Bott (left) and Ferdinand Piëch at Le Mans in June 1971. (Porsche Museum Archives)

RIGHT: Porsche's Peter Falk, Ferry Porsche, Anatole Lapine, Huschke von Hanstein, and his wife, Ursula, confer at Le Mans 1970. (Porsche Museum Archives)

through the woods. What a tale he must have told his wife!

Racing for the John Wyer Automotive Gulf 917 team, I was paired with Jo Siffert, a wonderful driver and friend. We won the opening race of 1971, the 1,000 km at Buenos Aires in January, which set the tone for the year, as Pedro Rodriguez, Jackie Oliver, Jo, and I simply dominated racing that season. What I learned from John Wyer was his profound attention to detail, and this certainly helped me in later

Professor Bott told us, "Every race must be a development of something; otherwise why do we go racing?"

years. The highlight of the year was sharing rides with these great drivers and learning so much from them. Can you imagine hurtling at 246 mph down the Mulsanne Straight in the night?

The 917 was simply an incredible race car. The sound of its air-cooled flat twelve alone raised the hackles on your neck. It was the car every racer in that era wanted to drive. In those days they refueled the car with churns of fuel, and we sat right in the middle of the fuel tank, with our feet ahead of the centerline of the wheels. We really must have been crazy, but we'd have done anything for a chance to drive this car.

People complained that rear vision wasn't good, but it didn't have to be, because you were always in front. It was the sports car of its era, and it's incredible, even today. The 917 meant more to me than any other car I drove.

Sadly, the tragic low point that year, which certainly jolted me into reality, was that both Pedro and Jo lost their lives, albeit not in Porsche cars. Racing truly was very dangerous. In October 1971 Richard Attwood and I won the Paris 1,000 km, the final race for the factory team, in a Gulf 917. What an amazing experience to have been part of that memorable era.

I drove various races in Porsche 935s for Georg Loos and the Kremer brothers in the mid 1970s. What tough cars they were to drive—you really had to roll up your sleeves and battle these cars to the end of the races. Without a doubt the highlight of my 935 times was winning the Lumberman's 500 driving with the brilliant Rolf Stommelen in the Andial Porsche. It was my first taste of working with Alvin Springer, who did so much for Porsche racing in IMSA for Porsche Cars North America over many seasons thereafter.

They provided superbly prepared motors for all those races when I competed for another remarkable man, Al Holbert. Co-driving the 924 Carrera GTR for Porsche at Le Mans in 1980, we were fifth overall at eight o'clock in the morning, but the Mulsanne Straight curse bit again, as it has for many engines over the years. Three point eight miles every lap, at full throttle, affected the lubrication of the bores, and we lost a cylinder. Running on three cylinders for eight hours, we finished the race, but around twelfth. The Martini Porsche 936 won the race.

I was on the verge of retiring in 1980, when I had a phone call from Steve O'Rourke, manager of the rock group Pink Floyd and owner of several vintage race cars, for 1981. He mentioned Lancia; I suggested BMW M1, so after a short discussion, that's the direction we went.

At our first race at Monza in the 1,000 km, Valentin Schaeffer, a Porsche Turbo engineer whom I knew very well, suggested I should drive the factory 936 at Le Mans. Steve O'Rourke released me from my contract, as he knew I had a great chance to win. I drove with Jacky Ickx in the Jules 936, and that began a Porsche relationship for me with that wonderful team. All we did was refuel, change tires and drivers, and keep out of trouble, and we passed the checkered flag first.

In my opinion 1981 was the beginning of the greatest era in the history of sports car racing. The trendsetting 956 was the car everyone wanted to drive, and Jacky Ickx, Jochen Mass, Vern Schuppan, and I were teamed together.

Porsche planned a horizontally opposed engine in a ground-effect car! *"No one has done this before,"* Dr. Helmuth Bott (head of research and development) pointed out, as I hesitated over signing my contract: *"But we have never been wrong before!"* So I signed!

The changes from the 956 to the 962 did not affect the handling much at all, but they allowed the 962 to be raced in the USA, albeit with a single turbo and based on the former 935-style engine. The 962 raced and won everywhere in the world. The World Endurance Championship (Group C) cars and, in our case, the Rothmans Team, used the engine that had been designed for Indianapolis (running on pump fuel, not methanol) with twin turbos and other technical advancements. Each year the improvements were minimal, with the engine increasing from a 2.4 liter with around 600 hp to 3.2 liters and over 800 hp in qualifying form by 1988. It seemed as though the speeds improved every month. Electronic modifications in the fuel delivery system improved economy, and the lessons learned aided the development of road cars. The engineers worked on the PDK transmission (*Doppelkupplung*, which shifted manually and automatically), which we then tested and raced. That caused us all some real frustrations, but when we complained, Professor Bott told us, "Every race must be a development of something; otherwise why do we go racing?"

In the USA we ran a single turbo and slightly higher ride height, but basically the same car as in Group C. But this car had a more violent, surging delivery of its 800 hp, allowing fabulous power slides. Those glorious years attracted millions of fans worldwide. It was mainly attributed to the Porsche philosophy of providing so many cars for sale at a moderate price. Without that backing, sports car racing would never have been as successful. I was honored to be a part of it all.

Derek Reginald Bell MBE is a British driver who raced for Porsche from 1971 through 1993. Bell won Le Mans five times, the Daytona 24 three times, and the World Sportscar Championship twice. He also raced in Formula One for the Ferrari, McLaren, Surtees, and Tecno teams.

RIGHT: Porsche celebrated Derek Bell's second World Sportscar Championship with this poster in 1986. (Porsche Museum Archives)

MILES C. COLLIER

AUTOMOTIVE RESTORATION IN THE TWENTY-FIRST CENTURY

Restoration projects have changed markedly from the practices of only a few decades ago. Even today, we still dream of discovering a fabulous old car covered in dust, gently moldering in a barn. The glass is yellowed, the paint scabrous with rust, the interior redolent with mildew and rodent droppings. Yet for all that, this imaginary car is totally complete, needing only the sympathetic touch of the restorer to be reborn as the exquisite mechanical confection it once was. Alas, today's restoration projects fall far from this romantic notion.

Over the last fifty years, car collecting has matured from a hobby involving inexpensive old cars; through antiquarianism, focusing as it does on makes, models, and history; to today's emphasis on connoisseurship. Consequently, as with other major collectibles, many historic automobiles have exploded in value. The high end of the market now prices significant cars comparably with important art, antiquities, antiques, and gemstones. Few if any significant cars are left to be discovered. Those that are uncovered, like as not, are so deteriorated as to be virtually unsalvageable. The typical twenty-first-century restoration of an important car involves the re-restoration of an already restored automobile rather than the restoration of an untouched artifact. The incidence of re-restorations is directly linked to the increasing knowledge and sophistication of collectors, the rising value of fine automobiles, and the publication of definitive monographs on individual makes and models.

An archival restoration is centered on archaeological thinking, founded on historical fidelity, and supported by extensive research.

Increasing market awareness is driving demand for "archival" restorations to replace "hobby" restoration.

An archival restoration is centered on archaeological thinking, founded on historical fidelity, and supported by extensive research. It strives to present the automobile as it was in period. Older restorations, as we shall see, rarely hold up to the demands of sophisticated modern collectors.

Today's archival restoration standards are much different from those of the past. The major developments behind these changing standards are today's higher intrinsic automobile values; a preservation ethos that is replacing an "always restore" mindset that resulted in countless fine, original cars being vandalized; and, finally, owners' emerging sense of responsibility for preserving cultural heritage. Yet many of today's great restored cars still display hobby restoration practices to some degree, either because they were restored long ago or because they include some amount of hobbyist thinking in their current presentation. Let's take a closer look at customary hobby restoration practices and contrast them to today's emerging, research-based, archival method.

First, those older restorations were performed at a time when important cars were worth a fraction of today's values. The restoration standards of the day reflected that lower value. If new, off-the-shelf, manufacturer's parts weren't available, it wasn't economical to re-create those needed parts and components to the identical specification of the manufacturer's originals. Restorers either accepted better-condition used pieces or substituted components made from scratch, down to the price that could then be justified. Alternatively, it was not uncommon to employ anything that would just "bolt up."

Likewise, the older restoration involved the use of construction techniques that were available to the owner, from either his own skill set or that of the local workshop. Consequently, older restorations tend to incorporate both errors in configuration (wrong or crudely made replica parts) and errors in technique (electric welding in lieu of gas welding, or wheel shaping of panels when beating metal was the original forming method). While it is true that such restoration deficiencies have little adverse effect on today's value of an important car (the very identity and provenance of the object accounting for ninety-plus percent of its value), sophisticated collectors, marque experts, and knowledgeable members of the public recognize such shortcomings. Most important, elite concours judges and other arbiters will invariably recognize such failings.

And so comes the modern re-restoration based on forensics: objective, historically based sources that expose details of original configuration, manufacturers' build processes, and customer use. Forensic sources are period photographs, sales brochures, engineering drawings, and maintenance and parts manuals. Just as much effort goes into understanding original manufacturing processes, product quality standards, shop floor methods, and authentic materials as on finding detailed specifications of the automobile itself. When comprehensively and sensitively employed, the archival approach presents a restored car with no feel of restoration about it. Despite its obvious lack of ordinary wear and tear or signs of aging, the automobile seems as if it was transported to the present by time machine.

Attention to authenticity is extremely costly both in prerestoration research and in execution. Archival restoration demands no-compromise historical exactitude. At the same time, such restoration demands acceptance of new aesthetic standards informed by the historic preservation movement. Archival restorations are much less jewel-like and less exquisitely detailed than hobby presentations.

Second, automobile restoration started as a hobby. It was focused on the pleasure of the hobbyist at the expense of the artifact, to the extent that countless wonderfully original old cars were unnecessarily restored. Thus began the convention of always intervening with an old car at the full restoration level rather than at a conservation level. In comparison, archival restorations recognize the essential historicity of the old car and celebrate

Porsche Type 917K, 1971, Courtesy of The Revs Institute for Automotive Research, Photograph © 2013 Peter Harholdt

Helmut Marko and Gijs van Lennep raced 917 number 019 three times in 1971, at no time as successfully as their 1970 24 Hours of Le Mans victory in another Team Porsche Salzburg 917K. Their 1970 Le Mans distance record of 5,335 km at an average speed of 222.3 kph was unbeaten for 39 years, until Audi's R15 TDi Plus achieved 5,410.71 km.

23
CIBIE
LIGHTS
RallyeBilt
OVERALLS

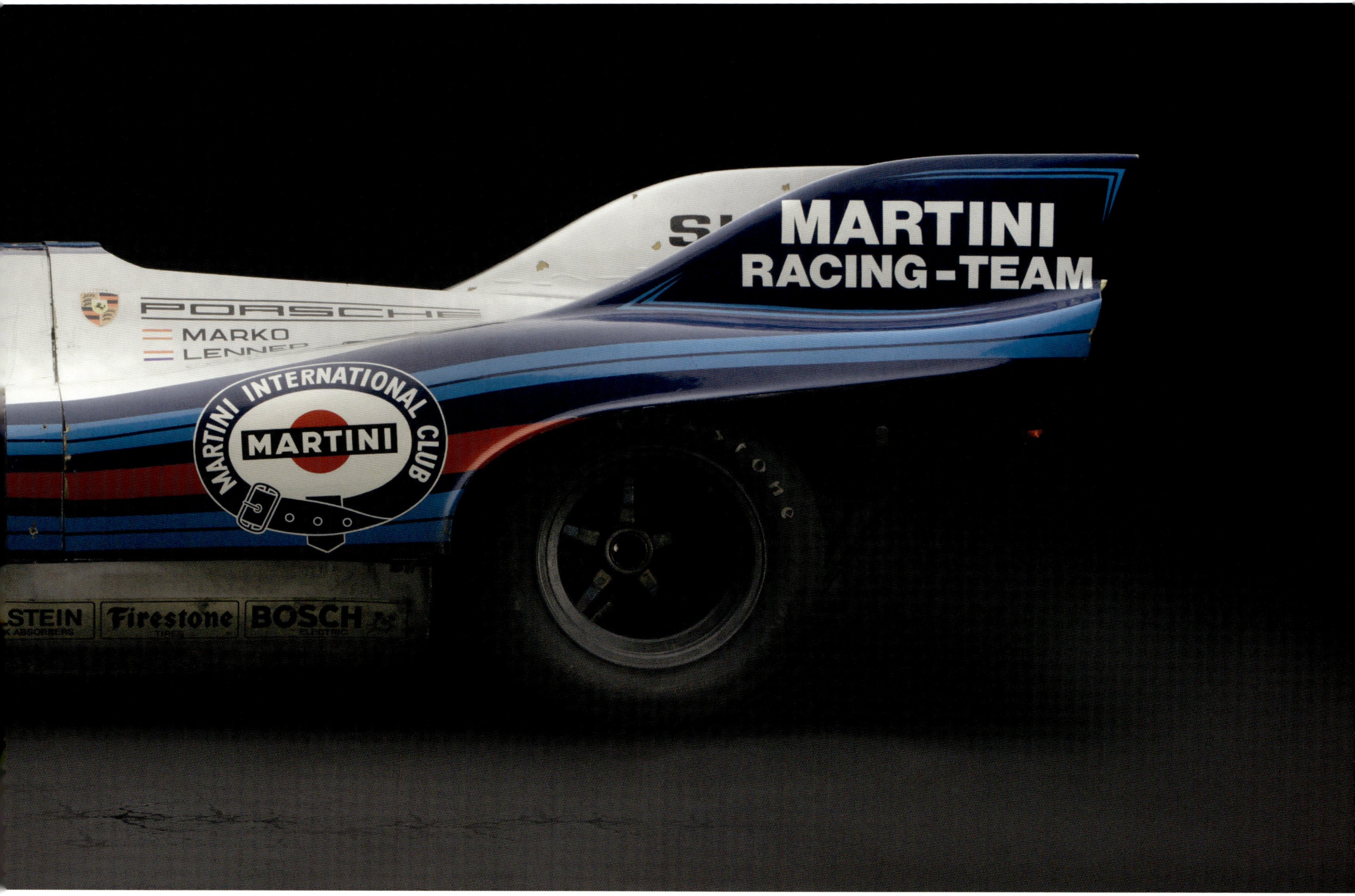
MARTINI
RACING-TEAM
PORSCHE
MARKO
MARTINI INTERNATIONAL CLUB
MARTINI
STEIN
Firestone
BOSCH

PORSCHE TYPE 917K

Porsche took racing seriously, but never so much as in 1969 at the Geneva Motor Show, when the company shocked the world with its most ambitious race car ever. The Type 917 represented a huge step that would catapult Porsche into domination of international racing.

Production Porsche cars were basically two-liter cars, and the company announced it would not build a larger road car to compete with Ferrari and Lamborghini. At that time V-8-powered Ford GT40s and Lola T70s were the main Group 4 competition. Porsche showed its serious intent by assembling the required twenty-five new five-liter 917s in just one month.

The 917's debut was unusual because Porsche had already committed to the 908. Under racing director Ferdinand Piëch (Ferry Porsche's nephew), Porsche AG had been introducing one new race car each year. It had never campaigned a large-displacement model, but anticipating competition from Ferrari and McLaren, Porsche stepped up decisively. Hans Mezger and his group were responsible for the design. Although it was not publicized, Volkswagen AG helped finance the 917; the success of air-cooled racing cars benefited VW as well. Air-cooled engines were potentially less powerful than liquid-cooled competition, but Porsche race cars weighed less without radiators. That was significant.

Historian Karl Ludvigsen quoted Piëch as saying that his uncle Ferry "let me carry on because our racing pleased him. He himself would never have done it, but … he happily tolerated the 917."

A flat-twelve engine was logical, as it was simply four cylinders more than the 908's and could be built nearly as compactly as a V-8. All the engine accessories

Porsche Type 917K, 1971, Courtesy of The Revs Institute for Automotive Research, Photographs © 2013 Peter Harholdt

Technical Notes

Engine Type: 4,907 cc, air cooled, horizontally opposed, fuel injected DOHC 12-cylinder
Horsepower: 600 hp at 8,400 rpm
Transmission: 5-speed manual
Top Speed: 223 mph
Wheelbase: 90.6 in.
Suspension: Front and rear, unequal-length upper and lower A-arms, coil springs
Weight: 1,760 pounds

were driven from the center of the crankshaft, including the glass-reinforced-plastic cooling fan. Engine cooling was aided by a remote tank that supplied an oil cooler. The first engines weighed just 530 pounds. The car's chassis was aluminum, saving even more weight.

The 917 showed its potential at the 1969 Le Mans trials, where Rolf Stommelen achieved 142.999 mph. The long-tailed versions were scarily unstable, baffling drivers and engineers, and after several incidents, the FIA banned the movable wings (elevons). Porsche competition manager Rico Steinemann convinced the Le Mans organizers that the car had to compete with elevons or Porsche would withdraw its entries. Vic Elford and Richard Attwood drove one of the 917s. Even with the elevons, Vic said it was "virtually undrivable." But "Quick Vic" set the fastest lap before the car was retired.

In 1970 England's John Wyer (JW Racing) was retained as Porsche's competition manager. His organization modified the 917's front end and developed a new shorter tail design based on the 917 Spyder. That solved the high-speed instability issues.

Wyer's toughest competition would come from the Austrian team, *Porsche Konstruktionen*, better known as Porsche Salzburg, a team fielded by Louise Piëch—Ferry Porsche's sister and Ferdinand Piëch's mother! Family ties ensured Porsche Salzburg received new 917 development parts as soon as, and sometimes before, Wyer's official factory team. In 1970 Porsche Salzburg won Le Mans, but JW's Gulf Wyer Porsche cars won every major race except the Sebring 12 Hours, where a new part failed and a new Ferrari 512S was victorious.

This car, 917-019, raced for Porsche Salzburg in 1970. The following year, 019 competed with Louise Piëch's Martini Racing Team. Once again the Piëch/Martini team won Le Mans. And the JW Gulf Porsche factory team won everywhere else, capturing the 1971 Sports Car Championship.

With its nine variants, the Porsche 917 remains one of the most powerful race cars of all time and starred in Steve McQueen's epic film *Le Mans*.

historic signs of use and aging by preserving them to the extent possible.

Finally, the biggest change has come in owner psychology. As hobby fodder, old cars gave collectors a canvas on which to express their nostalgic impulses or to embellish through nonprototypical polishing, unrealistically high standards of workmanship, and aesthetic improvements. By contrast, modern archival restoration places the welfare of the object above the ego needs of the collector. Formerly, restorations strove to show the historic artifact as exquisite, lapidary perfection. Today, archival restoration strives to present the artifact as historical narrative. While all narrative is subject to error in telling or in interpretation, the artifact as narrative has the power to return us to our historical antecedents. It has the power to make the past current and the abstract concrete.

Retired businessman, practicing artist, and authority on historic automobiles, Miles C. Collier has been involved in bringing scholarly attention to the automobile as social change agent and as museum-quality "worthy object." He is the founder of a biennial symposium that attracts car collectors from around the world as well as the Stanford Revs Project at Stanford University.

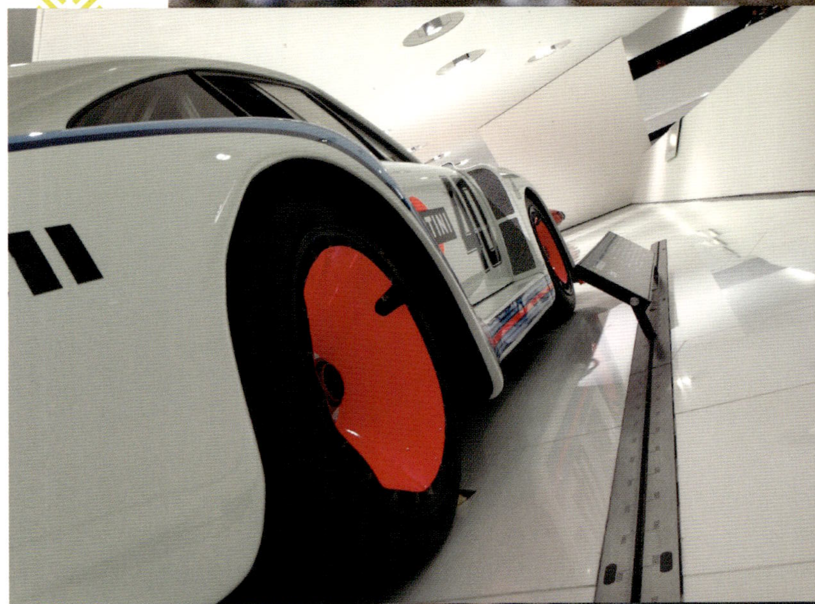

LEFT: Porsche Type 935 "Baby," 1977, in the Porsche Museum. Photograph © 2013 Barbara Wiedemann

ABOVE: Porsche Type 935 "Baby," racing at the Norisring in Nüremberg, Germany. (Porsche Museum Archives)

RIGHT: Porsche Type 935 "Baby," 1977. Courtesy of the Porsche Museum. Photograph © 2013 Porsche Museum

Technical Notes

Engine Type: 1,425 cc, air cooled, turbocharged, fuel injected, horizontally opposed, DOHC 6-cylinder
Horsepower: 380 hp at 8,200 rpm
Transmission: 5-speed manual transaxle
Top Speed: 168 mph
Wheelbase: 89.4 in.
Suspension: Independent front MacPherson struts, lower wishbones, adjustable antiroll bar; semitrailing rear arms and coil-over springs/shocks
Weight: 1,710 pounds

PORSCHE TYPE 935 "BABY"

Porsche was one of the few sports car companies to successfully "flip" a racing formula and win. In 1976 it introduced Type 935, the racing version of the 911 Turbo, to contest the FIA Group 5 Class. For 1977 the factory sold 935s to racing clients contesting the World Championship of Makes, the IMSA GT Series, and the German Racing Championship (DRM).

A 935 finished first overall at Le Mans in 1979; 935s won Sebring and Daytona and the Nürburgring. They were virtually unbeatable until the FIA changed the class rules in 1982, ending the 935's dominance.

Porsche built great race cars and creatively interpreted the rules. Engineer Norbert Singer found a loophole that let Porsche remove the 911's protuberant headlights and redesign the fenders to incorporate a flat nose, with headlights relocated to the front spoiler. This became a signature "look" for racing 935s. The flat nose and a winged rear were aerodynamically efficient.

In 1977, when 935s dominated Division I, German National Television announced that it would cover only the Division II race at the Norisring, where BMW and Ford were battling. With its larger displacement, the 935 was ineligible, but the formula also permitted a smaller 1,425 cc engine and a lower weight of 1,700 pounds. Porsche engineers built a special turbocharged engine for what would be the lightest 911 ever, weighing in at a featherweight 1,650 pounds. Much of its steel chassis was replaced with a tubular aluminum space-frame. The gas pedal and gear lever were titanium. The experimental 935 was nicknamed "Baby," and it was so light that the engineers had to add weight to meet the minimum.

At the Norisring the "Baby" and driver Jacky Ickx suffered from record heat. Porsche skipped the next round and prepared for the DRM support race at the 1977 German Grand Prix, where Ickx beat a Ford Escort and set the fastest lap. Team manager Manfred Janke told historian Karl Ludvigsen, "This car was built to prove one thing: that we are better than Fords and BMWs which are in this class in German races." Having proven its point, Porsche retired the victorious 935/2.0 "Baby."

Art Sports
the 0123
team 0123
PORSCHE
Art Corp
DUNLOP
SSR
B·B
53

LEFT: Porsche Type 962C, 1990, at the 24 Hours of Le Mans race in 1991. (Porsche Museum Archives); **Porsche Type 962C, 1990**, Courtesy of the William E. (Chip) Connor Collection, Photograph © 2013 Michael Furman

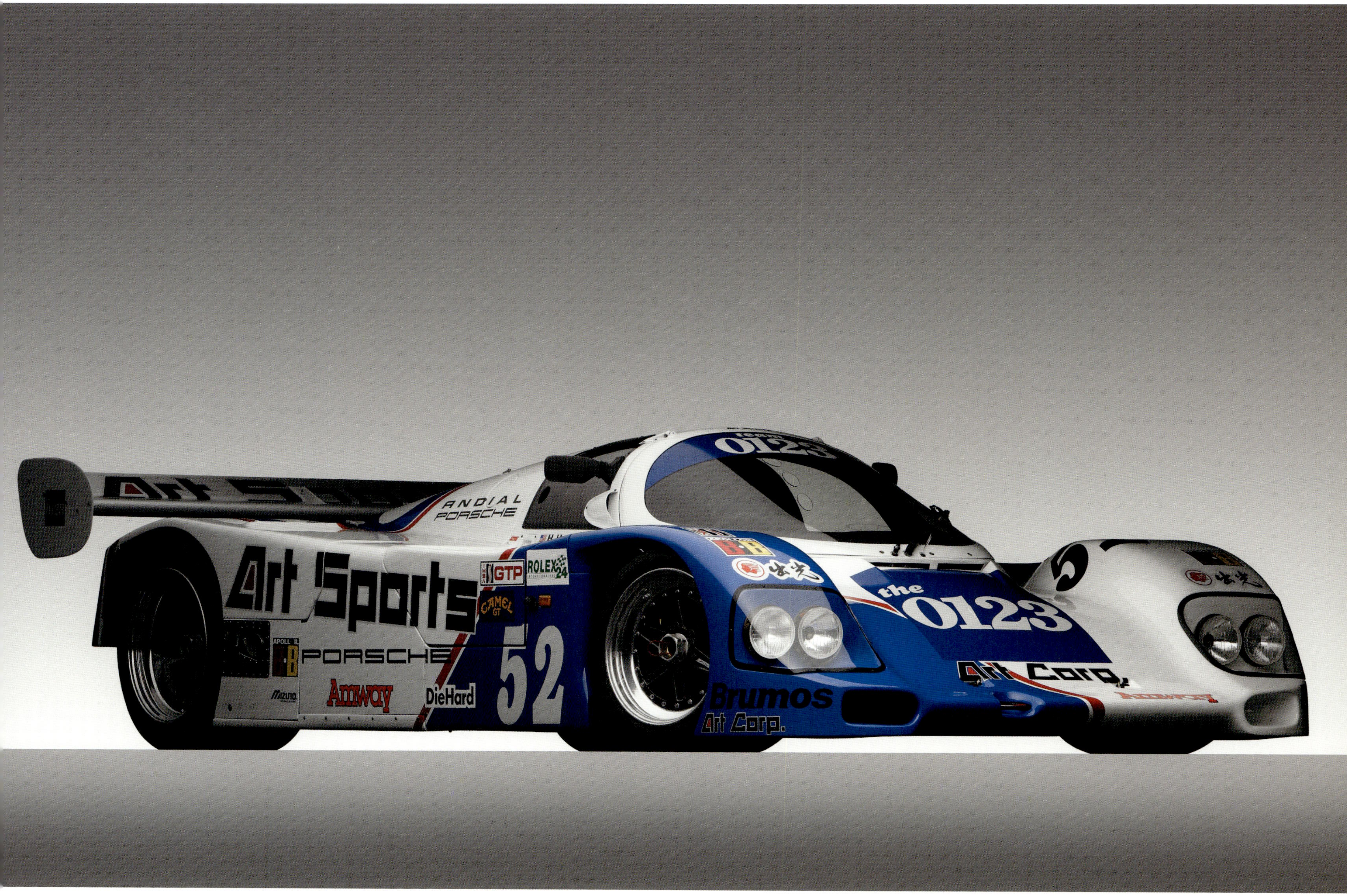

Porsche Type 962C, 1990,
Courtesy of the William E. (Chip) Connor Collection, Photographs © 2013 Michael Furman

Technical Notes

Engine Type: 3,200 cc, air cooled, twin turbocharged, fuel injected, horizontally opposed, 24-valve, DOHC 6-cylinder

Horsepower: 720 to 750 hp at 7,800 rpm

Transmission: 5-speed manual transaxle

Top Speed: estimated 215 mph

Wheelbase: 104.3 in.

Suspension: Independent front wishbones with coil-over springs/shocks; independent rear pushrods with coil-over springs/shocks

Weight: 1,984 pounds

PORSCHE TYPE 962C

In 1981 Porsche unveiled a new race car design aimed at capturing the FIA World Championship of Makes. The 956/962 Sports-Prototype racer was Porsche's first aluminum alloy monocoque design (in which the skin provides the main structural support). It was powered by several variations of the turbocharged flat six. The design took aerodynamic downforce to new levels with greatly improved road holding.

The Porsche 962, introduced in 1984, was an evolution of the Porsche 956, with the chassis lengthened slightly to place the driver's feet behind the front axle for crash safety. A steel roll cage was incorporated into the lightweight aluminum chassis. Some 962s were equipped with Porsche's PDK or double-clutch semiautomatic gearbox. Faster shifting than a manual gearbox, the PDK-type transmission would eventually be adapted for passenger car use.

One month after its debut in the Silverstone six-hour race in May 1982, Porsche 956 placed first, second, and third at the grueling 24 Hours of Le Mans. Porsche went on to win the coveted World Championship of Makes and the Driver's World Championship that year. In all, Porsche 956/962 models won ten World Championships and fifty National Championship races, and the 962 was still winning races ten years after its debut, making it Porsche's most successful racing car.

This example, chassis 146, competed with the Vern Schuppan Racing Team. Driven by many noted drivers, including Hurley Haywood, Wayne Taylor, James Weaver, Roland Ratzenberger, and Scott Brayton, it competed twice at the 24 Hours of Le Mans and raced at the 1992 Daytona 24-hour race. It is presented in 1991 Le Mans livery.

MICHAEL MAUER

STYLE PORSCHE: FROM THE PAST TO THE FUTURE

The Porsche success story is based on decades of experience in automobile manufacture and motorsports. On June 8, 1948, the first street-ready prototype of the Porsche 356 received its official approval for use on public roadways. Since that date Porsche has built over 1.8 million sports cars, all of which share the realization of Ferry Porsche's vision of "driving in its most beautiful form." The historic roots of Porsche AG, the tradition and values of the company, are today as important a part of the brand and company image as the Porsche attributes of technology, design, and performance dynamism.

Like no other automotive manufacturer, Porsche moves in widely differing categories, reconciling apparent contradictions such as sportiness and everyday practicality, tradition and innovation, exclusivity and social acceptance, and design and functionality. For us as Style Porsche designers, this complex brand identity is a special challenge, since our vehicle designs must reflect all of this. Today's Porsche design represents the continuation of an impressive tradition, which we carry into the future in combination with forward-looking innovations and careful doses of modernity.

This is particularly evident with the Porsche 911, which celebrates its fiftieth anniversary this year. Launched in 1963 as a successor to the Porsche 356, the 911 has become the heart of the brand. Few other automobiles in the world can look back on such a long tradition and such continuity. The Porsche 911 possesses a unique iconography

Porsche Type 901 Prototype, 1963,
Courtesy of Don and Diane Meluzio,
Photograph © 2013 Michael Furman

that has entered the collective memory and consciousness, taking on a symbolic identity as few other automobiles ever have. People take time to perceive new things, accept them, and ultimately want them. If the product is already obsolete by the time that happens, it cannot become an icon.

The 911 has succeeded in this during the course of five decades of evolution, making it a benchmark among sports cars, both for its performance and for its timeless design. Indeed, the design is undoubtedly a vital element of the 911 mythos. Many consider it to have the "ideal dimensions of the sports car," as a design critic once put it. Without question, with the 911 Ferdinand Alexander Porsche scored a major success. His design gave birth to a design culture that continues to inform our sports cars.

Throughout its seven model generations, the 911 has always been modern but never trendy—the mark of a true classic. We try to remain true to this principle in designing current and future Porsche sports cars, seeking to distill a design history whose roots go back to the 1930s. With the Volkswagen Beetle, the Porsche Design Bureau brought streamlining to the masses. The Type 64 of 1939 defined the original Porsche design idiom, entering production in 1948 as the Porsche 356. Yet with all due respect for the past, as designers we must not lose sight of the future, and we are too interested in the new for that to ever happen. The solution lies in the contemporary refinement of the form. This seems simple enough at first, but it is a great challenge for designers.

The silhouette of a 911 is quickly sketched, but the details are very difficult to capture; the design reacts extremely sensitively to even the smallest changes, perhaps because the basic form is so well proportioned.

In designing the current 911, we therefore first took a close look at the design history of

the series to find turning points. The first big change to the 911 took place in 1993, when the 993 model switched from convex to concave lines. The 996 generation went from air to water cooling and changed the headlights as well as the overall proportions. The very soft lines of the 996 were tightened up a bit in 2004 for the 997 model. On the current 911 (internally called the 991), this analysis of the predecessor models resulted in a longer wheelbase, wider track, and lower roofline. These changed proportions were taken as a basis to revisit the headlights and taillights, front mask, and rear window. The interplay of these areas gives the 991 its characteristic visual message, which is so important for the overall Porsche brand identity since the 911 is also the central point of reference for all other Porsche series. From the Cayenne to the Panamera, every Porsche is the most sporting automobile in its category, and each one carries a piece of the 911 philosophy.

This applies not just to the exteriors of present and future Porsche cars, but also—perhaps especially—to the interior. Unlike in the past, today the interior of an automobile is a major purchase criterion, and as a manufacturer we must pay great attention to it. The Porsche brand message must come through clearly inside the car. It's more than just a matter of the classic round instruments or the left-hand ignition. Our interior philosophy infuses all of our model series and is characterized by specific brand style elements. One current example is the rising center console common to all Porsche models. Introduced in the 2003 Carrera GT, it holds all of the important controls and switches.

The interplay of shared exterior and interior elements in all of our models gives Porsche a strong brand and product identity that is unique in the industry. This provides us with a framework in which we can extend our

LEFT TO RIGHT: A young F. A. "Butzi" Porsche at Porsche in June 1957; brochure cover for the early Type 901 (911); Type 911 advertising poster; F. A. Porsche with colleagues sending birthday greetings; a 1965 Type 911 Coupe in New York in 1967. (Porsche Museum Archives)

Porsche Type 901 Prototype, 1963, Courtesy of Don and Diane Meluzio, Photographs © 2013 Michael Furman

model range without running the risk of watering down the Porsche brand, as is convincingly demonstrated by the success of the new generations of Boxster and Cayman, and of the Cayenne and Panamera. Upcoming models like the Macan and the 918 Spyder also carry one hundred percent Porsche design DNA. As designers we have the privilege of living with one foot in the future, since we are always working on ideas for the automobiles of tomorrow and the day after.

What direction is the design of future Porsche models headed in? I am quite confident that the 911, as the core of the Porsche brand, will remain the point of departure for our design strategy. The clearly defined elements of this strategy will make all new models instantly recognizable as Porsches. Yet change will remain a constant, for we will continue to define new formal elements while discarding others, so that our design idiom is continually renewed. There is much to look forward to!

As chief designer at Porsche, Michael Mauer is responsible for the design and styling of the company's automotive and industrial products. He came to Porsche with design and management experience gathered at Mercedes-Benz, Smart, Saab, and General Motors Europe.

The silhouette of a 911 is quickly sketched, but the details are very difficult to capture; the design reacts extremely sensitively to even the smallest changes, perhaps because the basic form is so well proportioned.

Porsche Type 901 Prototype, 1963, Courtesy of Don and Diane Meluzio, Photograph

Technical Notes

Engine Type: 1,991 cc, air cooled, horizontally opposed, SOHC 6-cylinder

Horsepower: 148 hp at 6,100 rpm

Transmission: 5-speed synchromesh manual

Top Speed: 120 mph

Wheelbase: 87.1 in.

Suspension: Independent front MacPherson telescopic shock strut with lower wishbones and longitudinal torsion bars; independent rear transverse torsion bars with triangulated trailing links

Weight: 2,380 pounds

By the late 1950s, Ferry Porsche knew the 356 Series had run its course. Its rounded body shape was considered passé, and the pushrod four-cylinder engine had reached the end of its development cycle. The Carrera four-cam motor was too complex and expensive for series production.

Ferry's son Butzi began new design work, respecting guidelines from his father. Reutter Coachworks and designer Erwin Komenda also worked on the new model.

PORSCHE TYPE 901 PROTOTYPE

Ferry Porsche insisted the design be evolutionary and modern. It would still be a 2+2, with folding rear seats and a new, more powerful six-cylinder engine.

The new 901, introduced at the 1963 Frankfurt Motor Show, was more contemporary-looking than its predecessor. The all-steel unit body and chassis were rigid, and its independent front suspension was state of the art. The independent rear axle had inner and outer U-joints, resisting the old swing-axle suspension's alarming tendency to tuck under in hard cornering. Ferdinand Piëch and Hans Tomala designed the new SOHC, air-cooled six with eight main bearings, a forged steel crank, and a pair of triple-choke Solex carburetors. The new engine was mounted behind the rear axle.

The 901 was an immediate hit in Frankfurt. Before official production the 901 designation was changed to 911 because of a legal claim by Peugeot. This "signal red" coupe is the only survivor of thirteen prototypes used to develop the 911. After it was used for wind tunnel testing and heating system evaluation, it was sold to Richard von Frankenberg, editor of *Christophorus*, Porsche's magazine.

This hand-built car has bare rocker panels, plain flat hubcaps, and VW Beetle-like twin exhausts. The body is subtly different from a later 911's. The side window defrosters never made it into production; other differences include a twin-dial instrument cluster, a long shift lever like the 356's, a 356 steering wheel, and a round fuel-filler door. The sunroof tilts and retracts for headroom—a feature that was not repeated.

ROBERT CUMBERFORD

ALL PORSCHE IS DIVIDED INTO THREE PARTS

When people speak about Porsche road cars, they are really talking about the fabulous fastback coupes that stretch in a single unbroken—if slightly bent—line from hand-built 1949 Austrian Type 356/2 Gmünd cars to today's 911s. There have been many other Porsche cars, of course, ranging from shoebox-shaped 914s to incredibly fast turbocharged-truck SUVs to bloated four-door sedans to front-engine, rear-drive 924s, 944s, and 964s to that ultimate Teutonic Corvette, the 928. None of those matter. To purists and public alike, the word "Porsche" evokes rear-engine fastback coupes, pure sports cars.

That instant mental image derives from a single baseline and a single person. The engineering baseline is Volkswagen, a project of the design office the brilliant but irascible Doctor-Engineer Ferdinand Porsche formed in 1931. Porsche had held numerous high-level German auto industry posts during his tumultuous career. He finally found stability in his own engineering consultancy, forever best known for creation of the "strength through joy" people's car—the Volkswagen—and Auto Union mid-engine racing cars. The single person responsible for the indelible look that characterizes "Porsche" was Erwin Komenda.

Komenda, once head of all body engineering for Daimler-Benz, was among the founding personnel for Porsche's consultancy. It was he who transmuted the ill-fated 1934 DeSoto Airflow coupe's design into the VW shape that endured virtually unchanged from 1938 to 2006. And it was he who engendered the beloved Porsche coupe shape, first in the form

Ferry Porsche, Ferdinand Porsche, and Erwin Komenda (from right) look over a new Type 356 body circa 1950 at the Reutter factory in west Stuttgart. (Porsche Museum Archives)

of a VW racing/rally car, the 60K10 built in three examples in 1938/9, then as the 356/2 coupes built in a rural sawmill-lumberyard at Gmünd, Austria, in the late 1940s.

It was Komenda who shaped the very first Porsche-designed Porsche-brand car (now in the Zuffenhausen Porsche Museum collection), the Type 356-001 roadster, a one-off design built principally to provide sporty transport for Porsche's son Ferdinand, known as "Ferry." That one car generated two of the three main branches of Porsche development, the roadgoing sports cars and the racing line. The third branch is the heteroclite collection of expedient designs mentioned above that are really irrelevant to the Porsche image, if not to the firm's financial well being.

The 356 roadster's engine was placed ahead of the rear axle, as in all Porsche racers descended from it, and in the late 1960s VW-Porsche 914 and today's Boxster and

Cayman. Had circumstances in the immediate post–World War II period been less severe in war-ravaged Austria, where the Porsche clan gathered in its ancestral home to wait out the war years and plan its future, that Auto Union–inspired mid-engine layout would likely have been followed. But in such straitened conditions, with limited tools, onerous currency controls, travel restrictions, and constant oversight from the occupying Allied powers, only pragmatic simplicity could hope to provide a way forward. The tiny Porsche team, its leader still in custody in France, its revenues minimal, had to imagine a car it could make with whatever was at hand.

The team achieved something little short of a miracle. In an old sawmill in a remote rural area, some two hundred survivors of a brutal war, engineers and skilled tradesmen displaced from their peacetime facilities in Stuttgart, Germany, working with limited tools, using recovered ex-military VW mechanical elements and living hand-to-mouth, managed to build and sell half a hundred plump little coupes that evolved into one of the most respected and most profitable automobile manufacturers in the world. Porsche has achieved no fewer than sixteen wins in the 24 Hours of Le Mans endurance race, seven of them consecutively. But the firm has also managed to win the extremely difficult Targa Florio race eleven times, the last of them with a production-based 911. Add victories in the classic Rallye Monte Carlo, the trackless Paris–Dakar rally, and thousands of regional and national championships attained with 356 and 911 derivations, and one sees why only those cars truly represent "Porsche" to the world.

The slightly awkward Ur-Porsche 356-001 has had a hard life. Seriously damaged in an accident long ago, it was rebuilt with a considerably changed aspect, so only period photos show exactly what it was like. The 356/2 coupe derivation built in the lumberyard was no beauty, either. It possessed the typical German bottom-of-body datum, above which all surfaces rolled inward from the perimeter, with ineffective little bumpers finishing the form front and rear, but it had amazingly good aerodynamics, determined by eyeball, not wind tunnels. When the coupe was transformed from hand-hammered aluminum panels into stamped steel skins at the Reutter coachworks in Stuttgart in the early 1950s, its contours were still by Erwin Komenda, but the great designer refined and improved his initial design, retaining every element of its charm and effacing its minor faults. In its entire production run, the only noticeable changes to the 356 were in the shape of its windshield and bumpers.

Efforts to replace the 356 began quite early, with Komenda producing four-view drawings, first shown in the middle 1950s and very similar to the final 911 design. Albrecht Goertz, who had a consulting contract with Porsche, did some designs, but they were dismissed by the family as "a Goertz, not a Porsche." Goertz's main contribution to Porsche, he later said, was in taking on the third-generation Ferdinand Porsche—known as "Butzi" in the family—as a design apprentice, getting him out of academia. His doting father brought him into the company as head of Styling, and he applied his industrial design sensibility to a flat-sided design that was more consumer product than dynamically automotive, the result being a notchback four-seater now in the Porsche Museum as the T7. Ultimately it was Ferry Porsche who made the final determination that the new car would remain a fastback coupe, even if there were no real rear seats.

The Porsche company was a family enterprise until quite recently, when it was taken over by the Volkswagen group, run by Ferdinand Piëch, once chief engineer for Porsche racing cars. A series of *Dynasty*– or *Dallas*–like squabbles between third-generation cousins—four Porsche sons and Piëch—caused Ferry Porsche to decree that family members would no longer work in the company. The iconic and long-lived 911's design is entirely credited to Butzi Porsche in hagiographic company lore; Komenda, who died in 1966, is sometimes clipped as though he were a Soviet-era nonperson excised from May Day viewing stand photos. Eliminating the bulbous transverse sections of the 356 was all to the good, but there is no way to attribute the 911 entirely to F. A. Porsche. It is quite plain that today's 911 profile remains pure Komenda, whose work ties together the first and the last Porsche fastback coupes.

Independent designer and automotive design critic Robert Cumberford drove to his job—designing Chevrolet Corvettes—at the GM Tech Center in the mid-1950s in his 1954 Porsche 356 coupe.

FROM RIGHT TO LEFT in an informal gathering of Porsche colleagues and family: Franz Xaver Reimspiess, Leopold Schmid, Erwin Komenda, Louise Piëch, Karl Rabe, and an unidentified gentleman in bowtie. (Porsche Museum Archives)

FROM LEFT TO RIGHT: Looking over plans at the Porsche construction office in August 1959: Leopold Schmidt, Erwin Komenda, and Karl Rabe. (Porsche Museum Archives)

1951 Ferry Porsche (second from left, front row) with his closest colleagues: front row from left: Erwin Komenda, Ferry Porsche, Karl Rabe, Franz Xaver Reimspiess; back row from left: Franz Sieberer, Emil Soukup, Leopold Schmid, Leopold Jäntschke, and Egon Forstner. (Porsche Museum Archives)

IROC Porsche Type 911 Carrera RSR, 1974, Courtesy of the William E. (Chip) Connor Collection, Photograph © 2013 Michael Furman; this car (far right) in the 1973–1974 International Race of Champions series at Riverside International Raceway in California. (Porsche Museum Archives)

HULME
CHAMPION

GOODYEAR
PORSCHE

IROC Porsche Type 911 Carrera RSR, 1974, Courtesy of the William E. (Chip) Connor Collection, Photographs © 2013 Michael Furman

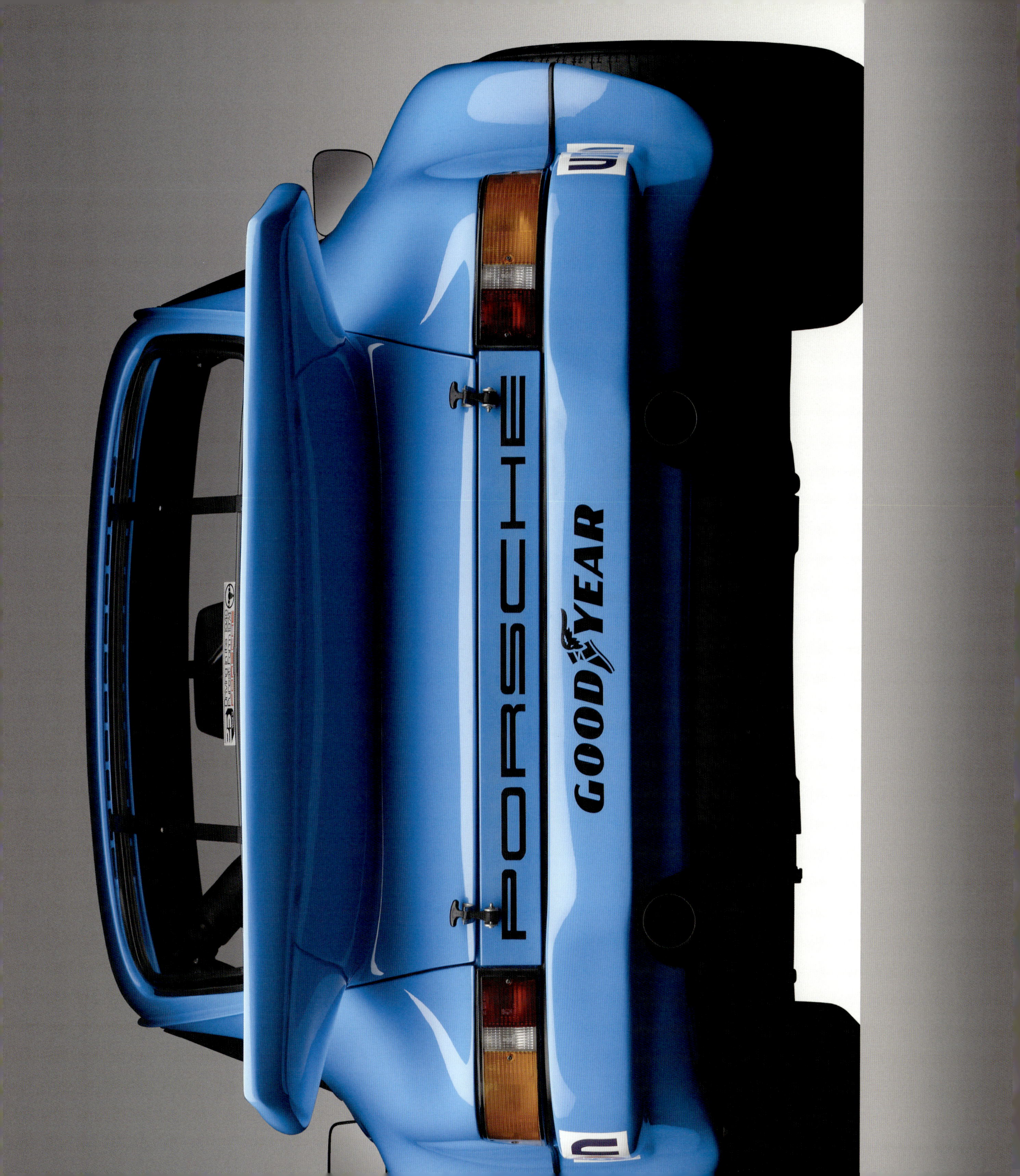
PORSCHE
GOODYEAR

RIGHT: IROC Porsche Type 911 Carrera RSR being driven by Formula One driver Denis Hulme at Riverside International Raceway in California. (Porsche Museum Archives)

LEFT: IROC Porsche Type 911 Carrera RSR, 1974, Courtesy of the William E. (Chip) Connor Collection, Photograph © 2013 Michael Furman

Technical Notes

Engine Type: 2,993 cc, air cooled, horizontally opposed, DOHC 6-cylinder

Horsepower: 316 hp at 8,000 rpm

Transmission: 4-speed synchromesh manual

Top Speed: 155 mph

Wheelbase: 89.4 in.

Suspension: Independent front transverse torsion bars with coil springs; independent rear multilink with coil springs

Weight: 1,980 pounds

The International Race of Champions series (IROC) was conceived by businessman and motorsports legend Roger Penske. Considered the perfect way to settle the perpetual argument over who was the best race driver, IROC was an "all-star" event featuring the top drivers of the day, competing in virtually identical cars. The model chosen for the inaugural series was the fast and reliable Porsche Type 911 Carrera RSR.

IROC PORSCHE TYPE 911 CARRERA RSR

With the 1973 2.8 Carrera RSR as a starting point, IROC racing cars received a larger engine, wider wheels, and full roll cages, along with larger rear spoilers and prominent "Porsche" logo bumpers. Penske ordered fifteen IROC Porsche cars (twelve race cars and three spares). They were finished in a rainbow of colors to facilitate TV viewing.

After initial reluctance to participate because of suspicions that longtime Penske driver Mark Donohue would receive preferential treatment, top drivers like Emerson Fittipaldi, Denny Hulme, Peter Revson, A. J. Foyt, Richard Petty, David Pearson, Roger McCluskey, and Donohue signed on. The plan was for three seventy-five-mile races at Riverside, Calif., and a ninety-five-mile finale at Daytona, Fla., in February 1974. Donohue won three of the four contests and was declared IROC champion.

This car, chassis number 0075, originally delivered in "Mexico blue," was raced in the 1974 IROC series by Hulme, McCluskey, and Pearson. Found with its original IROC engine, 0075 was restored by Hill and Vaughn, the Aase Brothers, and Charles Folkes. Purchased by Chip Connor in 1998, 0075 was campaigned in the 2002 Tour España and the 2003 Tour Auto, where it was co-driven by Danny Sullivan. A broken gearbox while leading the last race of the 2003 Tour Auto denied Connor and Sullivan an overall win.

D
CLASIC911
PORSCHE-BENSBERG

Porsche Type 911 Sport Classic Carrera, 2010, Courtesy of the Ingram Collection, Photographs © 2013 Michael Furman

PORSCHE
STUTTGART

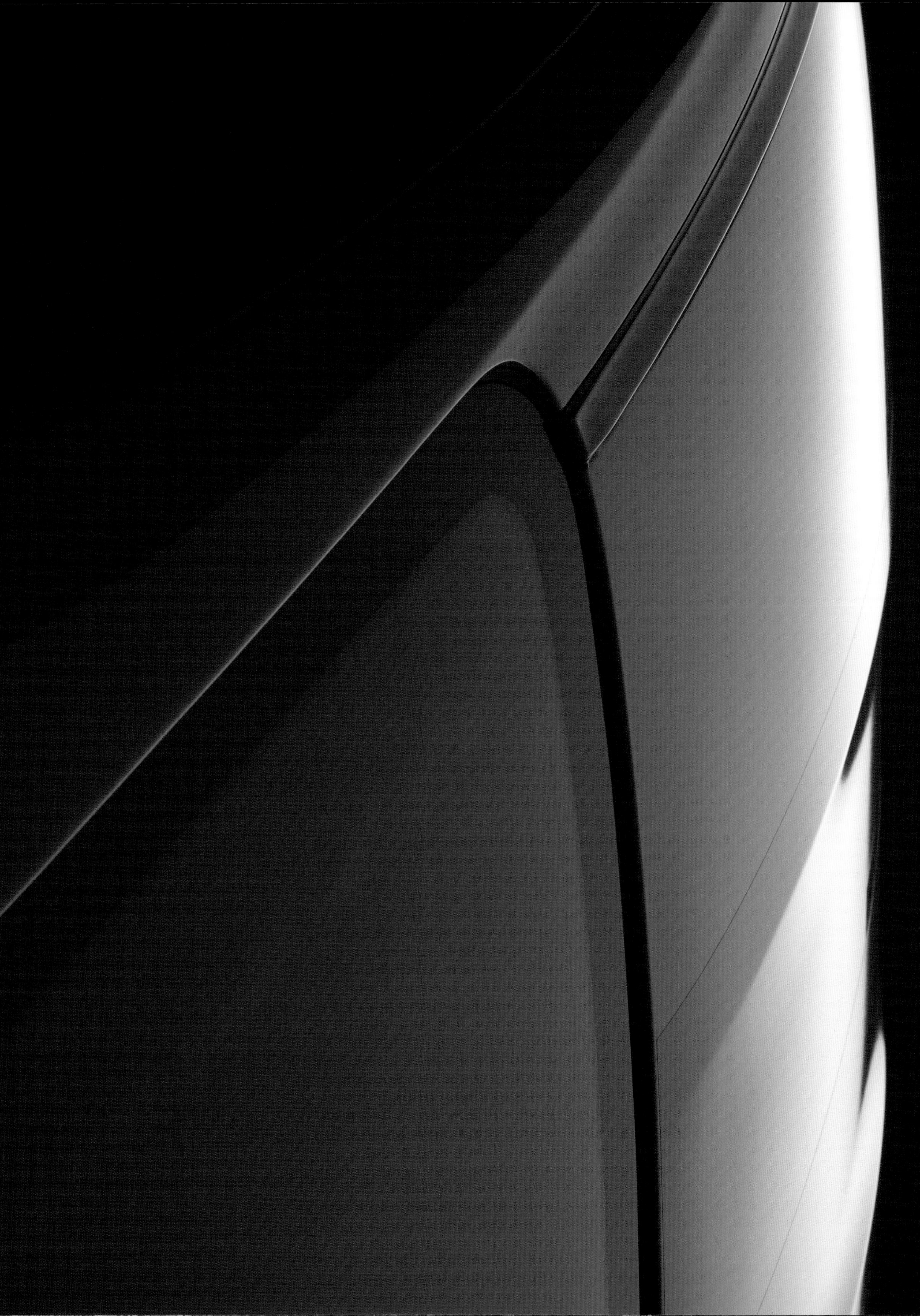

Porsche Type 911 Sport Classic Carrera, 2010, Courtesy of the Ingram Collection, Photographs © 2013 Michael Furman

Porsche Type 911 Sport Classic Carrera, 2010, Courtesy of the Ingram Collection, Photographs © 2013 Michael Furman

Technical Notes

Engine Type: 3,800 cc air cooled, horizontally opposed, DOHC 4-cylinder
Horsepower: 408 hp at 6,500 rpm
Transmission: 6-speed manual transaxle
Top Speed: estimated 191 mph
Wheelbase: 92.5 in.
Suspension: Independent front double wishbones; independent rear double wishbones with coil-over shock absorbers
Weight: estimated 3,131 pounds

PORSCHE TYPE 911 SPORT CLASSIC CARRERA

Talk about exclusivity: there are only four of these very special 911 Sport Classic coupes in the United States. The 911 Sport Classic was first seen at the 2009 IAA (Internationale Automobil-Ausstellung) Frankfurt Automobile Show. In 2010 Porsche AG in Germany offered European clients the chance to order one of these very limited production 911s. While the 911 Sport Classic is contemporary in every respect mechanically, it incorporates several "throwback" design features and color cues from classic Porsche cars of the past.

The exclusive Sport Classic coupes were equipped for the ultimate in grand touring, with many usually optional features fitted here as standard equipment. These included an engine that was specially tuned for 408 hp, race-inspired PCCB carbon-ceramic brakes, electronically controlled PASM sport suspension, a limited-slip differential, a prominent "ducktail" rear spoiler, nineteen-inch classic design Fuchs wheels, the Sport Chrono timing package, a navigation system, and adaptive sport seats with custom "espresso nature" (brown) leather.

The roof was altered with a subtle "double bubble" design reminiscent of the earlier 356 Abarth Carrera coupe, and the light gray/tan finish resembles a Porsche 356 color called "dolphin gray." It was almost as though a buyer had specified some of his or her favorite Porsche features over the years, and the factory obliged by building a custom 911 including all of them. This Sport Classic was purchased directly from the Porsche factory in Stuttgart. To date it has covered only about 150 miles.

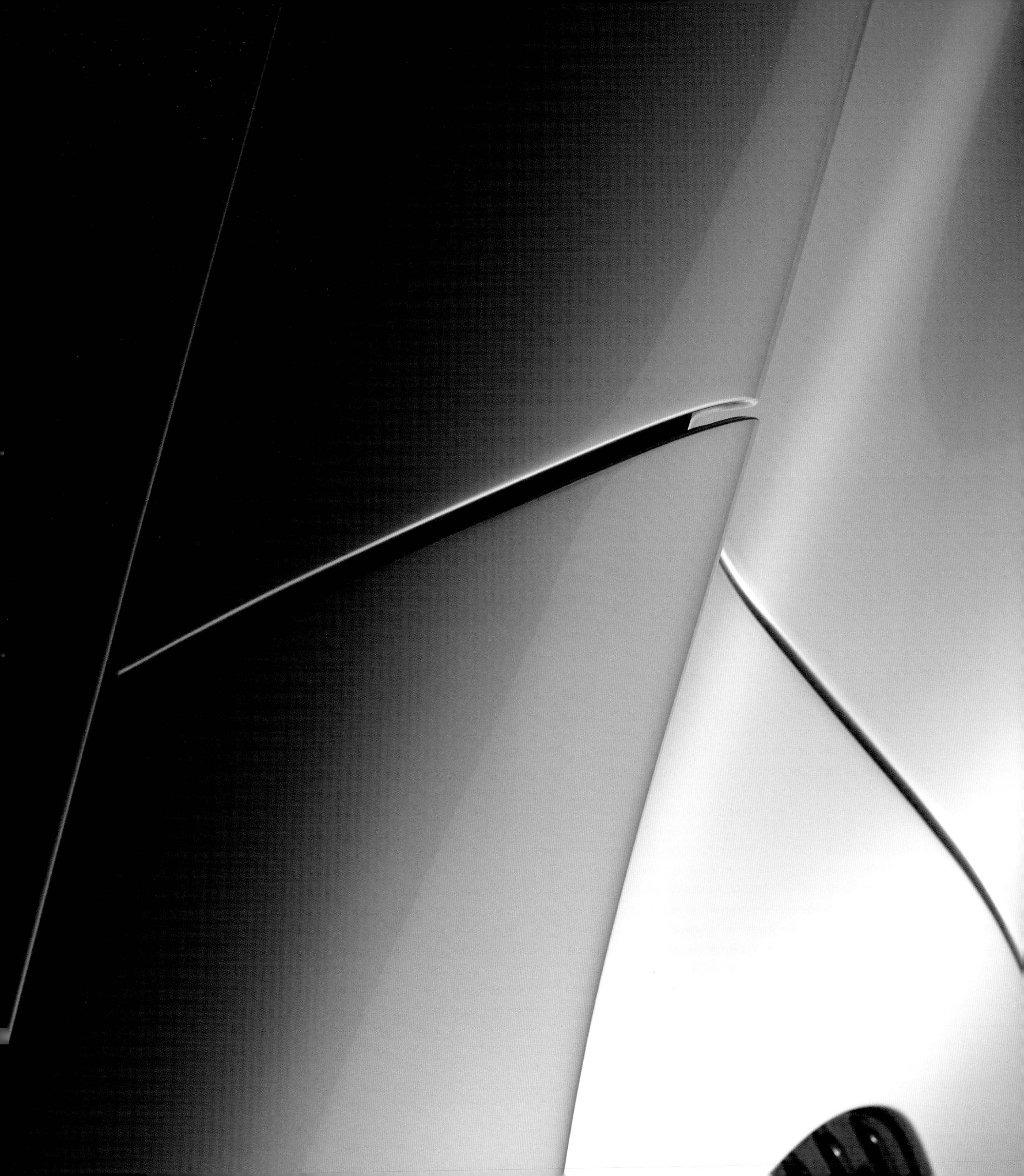

RANDY LEFFINGWELL **PORSCHE 911: A FINE SENSE OF STYLE**

Erwin Komenda worked with Professor Ferdinand Porsche beginning in 1929 and created car bodies for Auto Union and Volkswagen as well as for a prototype Cisitalia racing car. Komenda was Ferry Porsche's natural choice as his body designer when he felt the time had come to make his own automobile.

Komenda's job title was "body construction manager." He designed cars from the inside out, considering the elements that were necessary for the car, and then creating a body that surrounded them efficiently and attractively. Komenda believed large, gentle curves strengthened thin metal. He "floated" doors and front and rear deck lids away from panel edges to reduce metal stress and cracking. And he worked closely with modelers and panel beaters, who translated his elevation drawings into three-dimensional surfaces.

Ferry Porsche's son Ferdinand Alexander, known to family and co-workers as "Butzi" or F. A., grew up in his father's shops and became fascinated with automobile design. He watched modeler Heinrich Klie and body design engineer Gerhard Schröder. Ferry wanted front fenders with a pronounced shape so he could tell from the driver's seat where his front wheels were located. This led to the characteristic "stovepipe" fenders on the 911.

F. A. studied briefly at the Upper School for Art at Ulm, learning Bauhaus principles

Porsche Type 959, 1988, Courtesy of the Ralph Lauren Automobile Collection, Photograph © 2013 Michael Furman

Porsche Type 959, 1988. Courtesy of the Ralph Lauren Automobile Collection. Photographs © 2013 Michael Furman

where form followed function. Later when he supervised Klie, Schröder, and others as they developed the 911, they followed an industrial design approach using panel seams as the cut lines for doors and deck lids. In 1961 Ferry named F. A. head of "Styling." Komenda remained chief body engineer, in charge of making Styling's creations into cars that could be manufactured. The architectural modernity of Porsche's 911 emanated from F. A.'s Styling staff.

That classic 911, with its cylindrical front fenders, changed little for two decades, until Porsche chief executive officer Peter Schutz encouraged engineering chief Helmuth Bott to show the world the wider capabilities of the 911. Four-wheel drive fascinated Bott. The result was a model designated the Type 959.

"We did the show car," Dick Soderberg explained, "to see if anybody would be interested." Working for design chief Tony Lapine, Soderberg developed the appearance of the 959. Its shape resulted from practical considerations.

"We had to hold the door position," Soderberg said. "I think it was because of the structure of the car, the roof. [That meant] we had to terminate whatever other shapes we developed so they led into the door from front and rear. We were able to get [retain] the stovepipes, pull them down, get a little more angle on the headlamps. On the show car, we held the old hood from the original 911." The front of the car established form precedents that influenced later models, especially the Type 993. It was the rear of the car, however, that inspired copycats.

"We experimented with various configurations in the wind tunnel," Soderberg said. "The aerodynamicist was Herman Wurst, and he asked me if I could live with this shape if we do this sort of 'breadbasket' thing."

Porsche manufactured the 959 from 1986 through 1988, assembling 337 cars, including development prototypes, preproduction test models, and customer cars. They offered Sport models with a full roll cage and durable cloth interiors, and Touring versions featuring full leather interior. With electronically controlled "active" all-wheel drive, traction control, and advanced braking and suspension, it was the first of the supercars. Porsche lost money on each one, but the world's automakers lined up at Weissach for engineering and design advice.

As the last 959s left Zuffenhausen, management changes rocked Porsche: Helmuth Bott (chief engineer) retired, replaced by Ulrich Bez; Tony Lapine left, and Porsche hired Harm Lagaaij; and Peter Schutz vacated his office for Arno Bohn. Design chief Harm Lagaaij initiated a contest among his staff to design a show car for the 1989 Frankfurt exhibition. Steve Murkett had joined Porsche in time to collaborate with Soderberg on developing the production 959; he was intrigued with dune buggies and off-road vehicles, and his radical concept won.

"The idea with the Panamericana," Mur-

It was the first of the supercars. Porsche lost money on each one, but the world's automakers lined up at Weissach for engineering and design advice.

Porsche Type 959, 1988, Courtesy of the Ralph Lauren Automobile Collection, Photographs © 2013 Michael Furman

Drivers Al Hajri and Spiller won the "Pharaoh Rally" in a Porsche Type 959 in 1985. (Porsche Museum Archives)

Technical Notes

Engine Type: 2,849 cc, air cooled, horizontally opposed, twin turbo DOHC 6-cylinder

Horsepower: 450 hp at 6,500 rpm

Transmission: 6-speed manual PDK

Top Speed: 195 mph

Wheelbase: 89.4 in.

Suspension: Independent upper and lower double wishbones, front and rear, with Bilstein coil-over shock absorbers

Weight: 3,200 pounds

PORSCHE TYPE 959

One of the fastest supercars of its era, Porsche's all-wheel-drive 959 was developed to compete in FIA Group B racing. Although the Fédération Internationale de l'Automobile later dropped Group B, Porsche AG marketed these highly focused racing/rally models to offer clients a state-of-the-art race car for the street and to recoup this car's high development costs.

Its air-cooled flat-six engine upheld Porsche tradition, but it had more efficient water-cooled twenty-four-valve heads and titanium connecting rods. A detuned version of the 1981 Le Mans-winning Porsche 936, it was refined from lessons learned with the 956/962 models that won Le Mans every year from 1982 to 1987. The 959's lightweight composite plastic, Kevlar, and aluminum body had a relatively low drag coefficient of 0.32 (a contemporary 911 was 0.39) and could go from 0 to 60 mph in just 3.6 seconds.

A computer-controlled six-speed transmission with a multiple disc clutch (known as PDK) provided options for driving in dry, wet, icy, or off-road conditions. Ride height was electronically controlled, and the ABS brakes were based on those of the racing 935/956/917s. Bridgestone run-flat tires on magnesium wheels, with tire pressure monitors, eliminated the need for a spare.

The original selling price was about $295,000, but speculation and scarcity (only about 300 were built) drove prices up. Unlike most thinly disguised racing cars for the street, 959s were available fully equipped with such niceties as power windows and mirrors, heated seats, and a complete HVAC system.

Porsche 959s finished first, second, and sixth in the grueling Paris-Dakar off-road race in 1986. Production 959s were not initially legal for U.S. importation, but car racer and restorer Bruce Canepa of Mill Valley, California, pioneered the work needed for U.S. DOT/EPA approval. His company legalized many 959s for road driving, including this example.

kett explained, "was an open-wheeled vehicle, like an enduro motorcycle [off-road racing motorcycle]. Absolutely minimal, just the necessary skin to cover this shell; the wheels would stand free, and the car would stand way up on its four-wheel drive." Murkett conceived his car with "bladelike fenders, offset just like an enduro motorcycle. We were trying to get a Baja-buggy look."

From March to May, Murkett worked with modelers realizing his design. Lagaaij demanded a running show car, so Murkett started with a 964 Carrera 4 shell. They sliced away everything extraneous and applied clay right onto the steel.

"At the end of the project, when we presented it to Arno Bohn, he said, 'Wait a minute! Porsche doesn't produce an off-road car. We have a four-wheel-drive sports car, and we're trying to promote the idea that you need four-wheel drive on the road.' "

Murkett's team lowered the car. Flowing fenders enveloped the wheels. "Had we known," Murkett said, "the car could have

been designed a little bit more with that 'on-road' look in mind." They finished it in August. Murkett's off-roader had not required mufflers, so he left little space for them. "Exhaust went out of the engine through this collector and straight out the side of the car," he said. "The car had an amazing sound."

"And then, of course, this zip-on-zip-off roof system," Murkett said. "A crazy idea back then was to create a roll cage that didn't go over the roof, and instead it went around the roof. It protected your forehead. Then we'll use that roll cage, using a zipper, to make a cover for the car. One hobby I shared with my father was boating, and this came from my boating fascination, zips on covers. Two zips and you had the roof off."

Porsche showed the Panamericana at Frankfurt and Tokyo. When it returned, Lagaaij and Murkett slightly modified the body to improve handling. "Harm [Lagaaij] and I had this habit on a Friday afternoon when it got quiet here. 'C'mon,' he'd say. 'Let's go out for a run in the Panamericana.' " The following March

ABOVE AND FAR LEFT: Porsche Panamericana Concept Car, 1989, Courtesy of the Porsche Museum, Photographs © 2013 Porsche Museum

IMMEDIATE LEFT: On September 19, 1989, the Porsche Panamericana show car was presented to Ferry Porsche for his eightieth birthday. He is pictured here with his four sons, from left to right: Hans-Peter, Gerhard Anton, F. A., and Wolfgang. (Porsche Museum Archives)

Porsche Panamericana Concept Car, 1989, Courtesy of the Porsche Museum, Photograph © 2013 Porsche Museum

Lagaaij presented the car to Ferry Porsche for his eightieth birthday.

"There was this sense," Murkett said, "that if a car didn't look like an old 911, it wasn't a Porsche anymore." His colleagues took the Panamericana as permission to let go of the stigma. The show car created a new design language for Porsche.

A decade later CEO Wendelin Wiedeking pushed Porsche to think outside the box again. Following intense research he directed Weissach engineers to develop an SUV; not surprisingly, Murkett earned the assignment to design its exterior. But Porsche had just won the 24 Hours of Le Mans in 1988 with a prototype GT1 that was as startling to look at as Murkett's Panamericana had been. When racing engineers sought approval to develop the next Le Mans prototype in carbon fiber, Wiedeking questioned how many 24-hour victories Porsche needed to cement its legacy. With Mercedes-Benz, Ferrari, McLaren, and recently resurrected Bugatti developing new supercars, Wiedeking made his opinion clear. As Harm Lagaaij recalled, "Wiedeking said, 'They're all bullshitters. Let's show them our way!' " But under Wiedeking show cars had to be feasible, to be vehicles Porsche could sell profitably.

"The decision for the Carrera GT to be an open car came from me," Lagaaij said. "Racers say, 'Oh, it has to be a closed car. Rigidity… ' But you cannot sell 1,500 cars in that price bracket if the owners cannot show themselves."

Those are the design chief's jobs, to understand the customer, to lead and inspire his staff with a word or a song lyric or a gesture. The car's mid-engine cover was similar to the open roof. Its new V-10 performed and sounded like a Formula One car. Enthusiasts were curious, and owners needed to show it off without dirtying their hands. Could an engine cover show? Reveal? Tease? "The name 'negligee' came from me, to cover the engine," Lagaaij said. "I like names that are analogies, metaphors, plays on words."

The show car proved an international hit. Lagaaij assigned Tony Hatter to bring the "dream" car into reality. Hatter had been the 993's chief stylist, absorbing design cues from Murkett's Panamericana and Soderberg's 959. Hatter also had designed the GT1 that won Le Mans.

"The Carrera GT was a continuation of everything we had learned in the GT1 chain of cars, because the decision to make the CGT was based on, 'OK, now we have a Le Mans prototype, which was an open prototype carbon fiber monocoque with this V-8 engine in the back,' " Hatter said. "The decision was, 'Now, let's make a sports car out of this for the road.'" It grew clear that they needed to change everything from the show car, whether it was safety, or packaging, or snow chain clearances, or cooling the engine. "And of course the thing had to have a roof, which our boys hadn't really thought about."

While Hatter accomplished a great deal of design work on computers, fine tuning took place on scale models. The show car had air-cooling openings behind the doors. But air traveling along the body side does not automatically turn ninety degrees and go into air intakes, at least not without projecting air scoops.

"Those openings at the back had to be so big that the car would look nowhere near like the show car," Hatter said. "So the cooling had to go to the front. When you look at the car, it was just a plumbing exercise." There were also air channels underneath the car, and the diffuser in the back where the air goes out.

"We did a lot of proposals for the roof. It had to cover the occupants. It had to stay on at 200 miles per hour. We had to have wind-up windows, and the top of the windows at the B-pillar had to meet the roof." They devised two removable panels that owners would stack in the front trunk. "There were lots of compromises, and then we'd have a meeting with my boss. He wasn't happy with me, Mr. Lagaaij, and often we had the show car there and he'd say, 'Well, this doesn't look like the show car!' "

Some things were impossible. Others were unavoidable. Hatter and Weissach's engineers lengthened the wheelbase considerably over the show car. Front cooling and the way air hit the nose and flowed to the side required "corners" on the car body. This necessitated widening the car's front end. The show car was sharp nosed. That didn't work for engine cooling.

"The engine compartment [beneath Lagaaij's 'negligee'] had to look good. The idea to do an engine frame completely in carbon fiber and Kevlar was brand new. There is an engine cradle that bolts onto the back of the bulkhead. If you put your hands underneath these beautifully formed carbon fiber pieces, you find it's not solid. It's a U section. Hidden in that U section are all the cables and wires our motor people love to have all over the engine. So we completely [hid] the cables from view in the engine compartment. Then it looked like a show car." As Hatter continued his overall design process, he moved on to full-scale models.

"Creating a car digitally is what people think happens. But you cannot get over having a clay model or a model in 3-D to walk around and appreciate. A digital model is either on a screen, or you project it onto a wall, but when the presentation is over, they switch everything off, and it's gone. A clay model is there for the visual subtleties. The chief can walk around it at night or the next day or a week later. Other people walk around it, make comments. They sink in, because you have to get a bit of distance from these things for a while, and other people's input is valid. You get too close to it," Hatter said.

"Every project we do, we learn a lot. And we think, 'OK, that was good,' or 'Let's not do that.' " Hatter presently is responsible for all sports cars within the design department. Style Porsche has engineers and designers working together. "It's great because we have a super engineering department and a heritage that we know and appreciate."

Michael Mauer is Hatter's current design chief. For 2013 the Porsche car family encom-

Porsche Panamericana Concept Car, 1989, Courtesy of the Porsche Museum, Photographs © 2013 Porsche Museum

Technical Notes

Engine Type: 3,162 cc, air cooled, horizontally opposed, DOHC 6-cylinder
Horsepower: 275 hp at 6,100 rpm
Transmission: 5-speed synchromesh manual
Top Speed: 162 mph
Wheelbase: 89.5 in.
Suspension: Independent front struts with lower wishbones, coil springs, stabilizer bar; independent rear with semitrailing arms, coil springs, stabilizer bar
Weight: 1,762 pounds

PORSCHE PANAMERICANA CONCEPT CAR

How about this for an exciting birthday present? When Ferry Porsche turned eighty, the factory presented him with this drivable design study based on the new Porsche all-wheel-drive Type 911 Carrera 4. Designed by Harm Lagaaij, it was inspired by a sketch from British designer Steve Murkett, completed in just a few months, and named for Porsche's successful efforts in La Carrera Panamericana, the long-distance Mexican road race. Initially shown at the 1989 International Motor Show in Frankfurt, it was a surprise hit; Porsche had not publicized it beforehand.

The Panamericana's horizontal roof structure influenced the subsequent development of the Type 911 Targa (a coupe with a liftoff roof section), and the show car's racy fastback roofline inspired the Porsche Boxster. An intriguing blend of angularity and compound curves, the design study is reminiscent of the Porsche 911 but at once distinctly modern. Its sleek composite outer skin is a melding of lightweight but strong carbon fiber and fiberglass.

The continuous roll bar, shaped like the railing of an ocean liner, was designed to protect the car's occupants in the unlikely event of a rollover. At the same time, it formed the Panamericana's distinctively low but bright and airy fastback roofline. A transparent fabric top was secured with a zipper, evoking a nautical theme. Both the top and the rear window were removable. The roof had five different settings from open to closed. The wide tires had small Porsche crest logos carved into the treads.

The Panamericana's aggressive styling shocked many viewers in Frankfurt, and Porsche historian Karl Ludvigsen wrote that it was "the most controversial concept car in the company's history." Some likened it to a modern dune buggy. The design study's shape pointed the way to possible future Porsche designs and was a preview of the Model 993.

Not widely seen outside Germany, now nearly twenty-five years old and better accepted than when it first appeared, the Panamericana remains contemporary looking, and in that sense it may be considered a tribute to the keen vision of its designers.

passed the 991, the Boxster, the Cayman, and the 918, and the Tiguan, Cayenne, and Panamera multipassenger vehicles. The family resemblance was carefully managed.

"Our design process," Mauer explained in an interview in late 2012, "has three major steps, and the first is to handle the proportions. A lot of companies underestimate the importance of this and then wonder why they can't fix it with styling. For us the design strategy and philosophy follow a pattern. That means brand identity, naturally. But product identity is an individual execution."

Adhering to patterns across vehicle lines and decades guaranteed the family resemblance between 959, Panamericana, and Carrera GT models. That same emphasis on "individual execution" has directed Porsche styling from the 356 to the 918.

Author and photojournalist Randy Leffingwell has worked for AutoWeek *magazine and several major daily newspapers, including the* Los Angeles Times, *where he wrote and photographed his first book,* American Muscle. *After writing a number of books, including several about Porsche, he became historical editor of* Excellence: The Magazine about Porsche. *He lives in Santa Barbara, California, and serves as a docent at the Mullin Automotive Museum in nearby Oxnard.*

Porsche Type 980 Carrera GT, 2005, Courtesy of the Ingram Collection, Photograph © 2013 Michael Furman

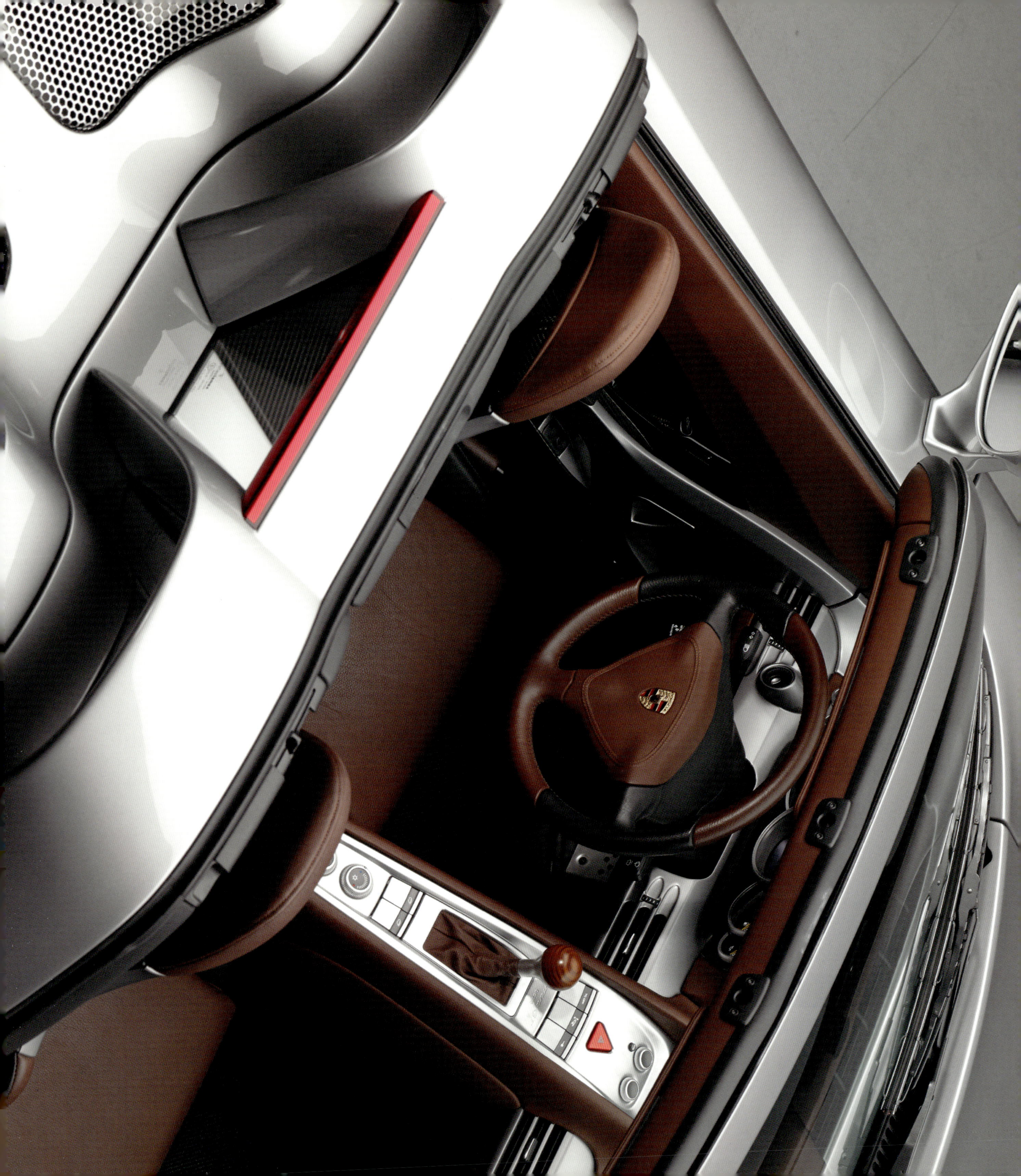

Porsche Type 980 Carrera GT, 2005, Courtesy of the Ingram Collection, Photographs © 2013 Michael Furman

Porsche Type 980 Carrera GT, 2005, Courtesy of the Ingram Collection, Photographs © 2013 Michael Furman

Technical Notes

Engine Type: 5,733 cc, air cooled, twin turbo DOHC V-10

Horsepower: 605 hp at 8,000 rpm

Transmission: 6-speed manual PDK

Top Speed: 205 mph

Wheelbase: 107.5 in.

Suspension: Independent upper and lower double wishbones, front and rear, with Bilstein coil-over shock absorbers

Weight: 3,043 pounds

Unveiled at the 2001 Paris Auto Show as the "Concept GT," the Carrera GT began as a design study for a Le Mans race car. Three years later, Porsche announced a production version with few changes. The Carrera GT's rigid unit construction, carbon-fiber-reinforced chassis tub, and lightweight carbon-fiber rear subframe cradle a Formula One–inspired V-10 engine.

PORSCHE TYPE 980 CARRERA GT

The GT's six-speed manual transaxle is mounted behind and below the engine. Its virtually fadeproof clutch contains four friction discs; its small size, coupled with a dry-sump oiling system, permits lower engine mounting for an optimally low center of gravity. The Carrera GT achieved 0 to 60 mph in 3.5 seconds.

Saving weight was a priority. The carbon-fiber seats weigh just twenty-three pounds each. The wheels are forged magnesium castings. The engine block is aluminum; the connecting rods are titanium; and the engine oil tank is an integral casting with the transaxle housing, saving even more weight. As befitting a supercar, the Carrera GT has enormous, composite, cross-drilled, lightweight ceramic brake rotors with six-piston one-piece aluminum calipers and four-channel ABS. Each wheel is attached with a single large nut that requires 400 pounds of torque for installation.

On September 7, 2005, on the Talladega Circuit, race driver David Donohue and NBC–TV *Tonight Show* host Jay Leno set closed-course standing start and flying kilometer and mile records in a Carrera GT. Leno was timed at over 182 mph; Donohue's best time was 198 mph. "This is where they separate the real men from the comedians," quipped Leno, whose standing-start 156.603 mph record for the mile in a Carrera GT, modified only with Michelin racing tires, attests to the GT's high-speed stability and handling.

Leno lost control for a millisecond at over 180 mph, spun around several times, and flat-spotted the tires. Zooming back into the pits, he jumped out of the GT and asked, "Did anybody see that deer?" After the tires were changed, Leno set a world's standing kilometer record in the same car.

DAN NEIL BEHIND THE WHEEL OF THE PORSCHE GT3 R HYBRID

It felt like a long morning at our hotel, adjacent to the Estoril racetrack near Lisbon, in November 2011. It was some sportsmen's resort—Penha Longa they called it, golf and haciendas, it might have once been a monastery—and the sun just wouldn't come on. I spent the night tangling my sheets, sleepless from jet lag and the prospect, in a few hours, of stepping into the Porsche Type 911 GT3 R Hybrid, the factory's advanced-technology racing prototype.

Not to be dramatic, but frankly, I was concerned. I wasn't in fear for my life: The GT3 R is quite solidly built, as you can see for yourself as it sits in the Museum's exhibition. I was concerned with failure. People outside the car world might not appreciate how big this moment might be for a journalist. This was a Porsche works race car, fresh from the plane after dominating a race in California, and it was beyond state of the art. Sitting in the passenger-seat space was a Kinetic Energy Recovery System, a KERS, about the size and shape of a microwave. The system's vacuum-encased, thirty-one-pound flywheel, spinning between 28,000 and 36,000 rpm, stored amperage mechanically, rather than electrochemically, like a battery.

What made this machine special was, in effect, that it worked like a Toyota Prius: electrical power derived under braking could be stored and then

Porsche Type 911 GT3 R Hybrid Race Car Prototype, 2010, Courtesy of the Porsche Museum, Photograph © 2013 Porsche Museum

Technical Notes

Engine Type: 4.0-liter, water cooled, horizontally opposed, fuel injected DOHC, 24 valves, 6 cylinders (470 hp) plus two electric motors (202 hp)

Horsepower: 672 total hp

Transmission: 6-speed sequential, all-wheel drive

Top Speed: 175 mph

Wheelbase: 92.5 in.

Suspension: Front and rear electronic height-adjustable suspension with coil-over shock absorbers

Weight: 2,866 pounds

used when accelerating. There the Prius comparison ends. The Porsche's system unleashed a veritable lightning bolt to the chassis: two hundred additional horsepower at the front wheels for a maximum of eight seconds, with an interposed software called torque vectoring actually steering the car by applying torque dynamically across the front wheels. Not just electro-turbo boost, if you will, but turbo steering. All this was in addition to the fire-breathing 470-horsepower flat six-cylinder gas engine in the back.

The GT3 R (2,860 pounds, 760 hp) accelerates from 0 to 195 mph in about fourteen seconds. Running in a demonstration class at a six-hour endurance race at Laguna Seca, this car finished three laps ahead of its would-be GT competitors. It stopped only three times for fuel, while the nearest rival was obliged to stop five times. This car represented not only millions of Euros in proprietary research, but also the sweat and toil of the many brilliant engineers milling around the hotel, looking apprehensive.

It would be a shame if anything happened to it.

The flywheel KERS device was developed by the British racing and aerospace concern Williams as part of Formula One's recent effort to be more socially relevant (it may surprise you to know racing is widely criticized by environmentalists around the world). And, for the record, F1's green push with KERS worked unusually well. The Porsche Type 911 GT3 R Hybrid is a direct result of that initiative. This technology will also be applied to Porsche's next production supercar, the rapturous 918 Spyder (except the street-legal 918 will use a lithium-based battery for power storage).

Indeed, car companies across the globe are incorporating KERS-like technology into their next-generation starships: the recently unveiled, million-Euro LaFerrari, the standard-bearer for the famed Maranello firm, has a rated output 963 horsepower, 163 hp (120 kW) of which is electrically produced; The coming-soon Acura NSX takes an approach similar to the Porsche, with torque-vectoring all-wheel drive. Under new hybrid-friendly rules from the Automobile Club de l'Ouest, Audi won the 2013 Le Mans with its KERS-equipped R18 e-tron quattro. In 2014 Porsche will challenge its VW Group sibling with its own hybrid machine, returning to the top class at Le Mans for the first time since 1998. This race will be the Battle of the Hybrid Titans.

But on that morning in Estoril, all I knew was that the GT3 R represented the dawn of a powerful performance technology, and that I was pretty close to being in over my head, driving-wise.

About midday, after some reconnaissance laps of the Circuito do Estoril in a conventional 911 Carrera S, I was belted into the GT3 R for a ten-lap stint. The production 911's steel and aluminum shell remained, but the interior was gutted and crisscrossed with the tubular-steel roll cage. The racing seat was hard and tight, and the suede-wrapped carbon-fiber steering wheel bristled with buttons and a center LCD display. A technician reached in and cinched down my five-point harness, squeezing the breath out of me (these belts worked themselves slack during my session).

If you're not used to it, the experience of driving a race car can be pretty claustrophobic. The helmet, the head-restraining HANS device, the fire suit, the harnesses, the seat, the learning curve. Adding to the sensory overload you'd expect of a world-beating sport racer, the GT3 R's soundtrack included the wild spiraling song of the flywheel as it spooled up during hard braking. It was deafening and actually quite unnerving.

My first concern was getting the car out of the pits without stalling; race cars with sequential gearboxes still have foot-pedal clutches to get moving in the pits, and these clutches are finicky. I managed to embark in good form. I began breathing again by about the second lap.

By my third lap I could tell something important was happening with the GT3 R. The electrical torque-vectoring steering effect was extraordinarily reassuring, almost uncanny. For one thing, the regenerative braking software—instantaneously adjusting each front wheel motor's "drag" to maximize stability—nulled out much of the squirminess that race cars exhibit under maximum braking, particularly on an uneven surface like Estoril's. Second, the torque-vectoring system's capacity to overdrive, to accelerate, the front outside wheel in a corner, while gently slowing up the inside front wheel, produces hugely athletic, intensely physical cornering responses. The car's steady-state lateral grip is around 2g—that's twice the pull of gravity—thanks in part to the car's vast rear wing and aerodynamic bodywork. But it was the car's transient grip, the way it changed direction with such enormous commitment and intention, that had me swearing and laughing in my helmet. Everything was happening faster than I had ever experienced—the track unraveling frantically in front of me like a video game—but I felt strangely in control.

I described it this way in the *Wall Street Journal*:

It's like falling into some sort of weird asphalt gravity well. There's very little drama in the steering, no unpleasant electrical pushback coming through the suede-wrapping steering wheel… The car blasts out of turns with a fierce, arcing inevitability, with its mountain of software helping to balance the car on the very knife-edge of tire adhesion.

It adds up to nothing less than a revolution in race-car dynamics, a virtually self-adjusting car, a dancing alloy algorithm of stunning pace and rip-your-head-off handling, an all-wheel-drive car with optimized traction control that knows the track as well as or better than the driver. Then it's just bang-bang-bang, up and down the paddle-shifted gears.

With huge relief and no small amount of sweaty nausea, I returned the GT3 R to the pits without any off-piste drama. Compared to the handful of global automotive media invited to drive, my lap times were pretty good, better than midpack. I had not disgraced myself. However, for reference, my lap times were eight seconds slower than the factory driver, Joerg Bergmeister.

Almost two years later, as I consider the car in residence at the Museum, I recognize it as a landmark in the sexualizing of hybrid technology. Since the second-generation Prius in 2003, hybrid cars have been fighting an image problem, the abstemious virtue-nerd cousins of real cars. But in the near future, the quickest, most powerful, most seductive sports cars on the planet will be partially electric, imbued with a computerized surety and controllability that makes more of the car's performance accessible to mere mortals. For lots of reasons—including, may I venture, product liability—radical sports cars of the future will have to be hybrids.

For those ten laps at Estoril, the GT3 R made me a driving hero. If it can do that for me, it can do that for anybody.

Dan Neil is the automotive critic for the Wall Street Journal *and winner of the Pulitzer Prize for Criticism.*

PORSCHE TYPE 911 GT3 R HYBRID PROTOTYPE RACE CAR

This experimental Porsche Type 911 anticipates the future of high-performance sports cars. Exactly 110 years after Dr. Ferdinand Porsche built his first hybrid-drive automobile, Porsche AG introduced a modern variation on this visionary concept. The GT3 R hybrid combines a flat-six Porsche conventional gasoline engine with a pair of electric motors. It develops race-winning power and respects the environment with a high level of performance.

Introduced at the 2010 Geneva Motor Show, this car has advanced technology that joins the work of Porsche engineers and the Williams Formula One Racing Team, whose KERS (kinetic energy recovery system) technology reuses energy that's stored in a spinning flywheel instead of a battery. This flywheel, attached to an electric motor/generator and lighter than a comparably powered battery pack, can be fully charged and discharged multiple times per minute without overheating, something a battery or capacitor cannot do.

When the GT3 R's hybrid system is activated under braking, the electric motors reverse their function and send current to the flywheel motor. The generator can rush a burst of power to the front wheels. Torque vectoring, controlled via buttons on the steering wheel, delivers power proportionally to the front wheel with the best traction.

Porsche entered the Prototype 911 GT3 R in several endurance races. It led the 2010 Nürburgring 24 Hours before retiring with an engine failure unrelated to the KERS system. In 2011 the GT3 R won a four-hour race at the Nürburgring.

In 2012 Porsche introduced the 911 GT3 R Hybrid 2.0, a lighter version of this car, with cooling system improvements and other new features. Porsche will soon introduce the Type 918 RSR, a roadgoing version of this advanced sports car.

LENDERS TO THE EXHIBITION

Automuseum Prototyp, Hamburg, Germany

Bowman Motors

Collection of Chad McQueen

Collection of the Joplin Family, Courtesy of the Rock and Roll Hall of Fame and Museum, Cleveland, Ohio

Don and Diane Meluzio

The Ingram Collection

The Porsche Collection of Ranson W. Webster

Porsche Museum, Stuttgart

Private Collection of Cameron Healy and Susan Snow

The Ralph Lauren Automobile Collection

The Revs Institute for Automotive Research

The William E. (Chip) Connor Collection

CONTRIBUTORS

Derek Bell

Miles C. Collier

Robert Cumberford

Ken Gross

Cameron Healy

Cam Ingram

Randy Leffingwell

Karl Ludvigsen

Pete Lyons

Michael Mauer

Denise McCluggage

Dan Neil

Jeff Zwart

PHOTO CREDITS

Michael Furman: cover, frontispiece, pages 22–23, 27, 28, 31, 36–41, 42–47, 62–63, 65, 111–115, 117, 120–123, 128, 130–134, 136–145, 146–149, 156–161

Peter Harholdt: frontispiece, pages 32–33, 35, 52–59, 61, 68–69, 70–71, 74, 76–78, 82–84, 102–107

Bruce Meyer: page 34

Porsche Museum: pages 150–151, 152, 154, 162–163

Porsche Museum Archives: pages 5, 9, 10, 12, 13, 14–15, 16, 17, 21, 24, 26, 27, 30, 34, 41, 50–51, 58, 59, 60, 64, 66, 67, 69, 73, 75, 80, 81, 85, 89, 92, 93, 94–95, 96–97, 99, 108–110, 117, 118–119, 125, 127, 128–129, 135, 149, 151, 155

Photographers Hamburg: cover, pages 14 16, 19

Cam Ingram: page 21

Jeff Zwart: page 48

RB/Redferns/Getty Images: page 52

Cameron Healy: page 79

Hal Thoms: page 80

Pete Lyons: pages 86–87, 88

Art Howard: pages 90–91

Barbara Wiedemann: page 108

ILLUSTRATED CHECKLIST

ABOVE: **Porsche Type 64 Berlin-Rom Racer**, 1938, Collection of the Automuseum Prototyp, Hamburg, Germany

BELOW: **Porsche Type 356 Gmünd Coupe**, 1949, Courtesy of the Ingram Collection

ABOVE: **Porsche Type 550 Prototype**, 1953, Courtesy of The Revs Institute for Automotive Research

BELOW: **Porsche Type 356 Speedster 1600 Super**, 1958, Collection of Chad McQueen

ILLUSTRATED CHECKLIST

ABOVE: **Porsche Type 718 RS60**, 1960, Bowman Motors

BELOW: **Porsche Type 356B 1600 Carrera GTL Abarth Coupe**, 1961, Courtesy of the Ingram Collection

ABOVE: **Porsche Type 804 Formula One**, 1962, The Porsche Collection of Ranson W. Webster

BELOW: **Porsche Type 356C Cabriolet**, 1965, Collection of the Joplin Family, Courtesy of the Rock and Roll Hall of Fame and Museum, Cleveland, Ohio

ILLUSTRATED CHECKLIST

ABOVE: **Porsche Type 356C Carrera 2 Coupe**, 1964, Courtesy of the Ingram Collection

BELOW: **Porsche Type 901 Prototype**, 1963, Courtesy of Don and Diane Meluzio

ABOVE: **Porsche Type 904/6 Prototype**, 1965, Private collection of Cameron Healy and Susan Snow

BELOW: **Porsche Type 908K Prototype**, 1968, Private collection of Cameron Healy and Susan Snow

ILLUSTRATED CHECKLIST

ABOVE: **Porsche Type 917 16-cylinder Spyder Prototype**, 1969, Courtesy of the Porsche Museum

BELOW: **Porsche Type 917K**, 1971, Courtesy of The Revs Institute for Automotive Research

ABOVE: **IROC Porsche Type 911 Carrera RSR**, 1974, Courtesy of the William E. (Chip) Connor Collection

BELOW: **Porsche Type 935 "Baby,"** 1977, Courtesy of the Porsche Museum

ILLUSTRATED CHECKLIST

ABOVE: **Porsche Type 959**, 1988, Courtesy of the Ralph Lauren Automobile Collection

BELOW: **Porsche Panamericana Concept Car**, 1989, Courtesy of the Porsche Museum

ABOVE: **Porsche Type 962C**, 1990, Courtesy of the William E. (Chip) Connor Collection

BELOW: **Porsche Type 980 Carrera GT**, 2005, Courtesy of the Ingram Collection

ILLUSTRATED CHECKLIST

ABOVE: **Porsche Type 911 Sport Classic Carrera**, 2010, Courtesy of the Ingram Collection

BELOW: **Porsche Type 911 GT3 R Hybrid Race Car Prototype**, 2010, Courtesy of the Porsche Museum

Note: Illustrations are indicated by page numbers in italic type.

Dedicated to Porsche designers—past and present—working under the guidance of:

Erwin Komenda

F.A. "Butzi" Porsche

Anatole "Tony" Lapine

Harm Lagaaij

Michael Mauer